SpringerBriefs in Plant Science

For further volumes:
http://www.springer.com/series/10080

Robert J. Dufault

Stalking the Wild Sweetgrass

Domestication and Horticulture of the Grass
Used in African-American Coiled Basketry

 Springer

Robert J. Dufault
Emeritus Professor of Horticulture
Department of Horticulture
Clemson University
Clemson, SC, USA

ISSN 2192-1229   ISSN 2192-1210 (electronic)
ISBN 978-1-4614-5902-6   ISBN 978-1-4614-5903-3 (eBook)
DOI 10.1007/978-1-4614-5903-3
Springer New York Heidelberg Dordrecht London

Library of Congress Control Number: 2012951316

Printed on acid-free paper

Springer is part of Springer Science+Business Media (www.springer.com)

This book is dedicated to the spirits of the enslaved Africans who brought the art of African-coiled basketry with them to America; almost three centuries later, the world still enjoys their gift to us. Also, this book is dedicated to Mary Jackson, an artist extraordinaire. Mary was my mentor and teacher on the art of sweetgrass basket making; I appreciate the wealth of knowledge and friendship she shared with me so I could develop a deeper appreciation for this ancient art.

Contents

1   **Introduction** 1

2   **The Beginnings of Change** 13
Sweetgrass Conference, 1988 13
The Findings of the Conference 17

3   **My Time to Get Involved** 21
What Were the Sweetgrass Basket Community Needs? 22

4   **Getting to the Grass Roots of Cultivation** 25
First Site of Large-Scale Sweetgrass Plantings: Palm Key Resort,
Ridgeland, SC 26

5   **Getting More Involved** 29
Low Country Sweetgrass Preservation Society 31

6   **Sweetgrass Utopia in the Southeast: Little St. Simons Island** 33
Little St. Simons Island History 33
An Amazing Sweetgrass Habitat 36

7   **Getting to Know the Basketmakers** 41
Mt. Pleasant Basketmakers' Association 41
Perceived Cultural Problems with Cultivated Sweetgrass 42

8   **Sweetgrass Culture Workshop: October 24, 1992** 43

9   **The Concept of Large-Scale Sweetgrass Plantations** 47
Grand-Scale Experiments and Perceived Problems 48

10 **Second Site of Large-Scale Sweetgrass Plantings:**
McLeod Plantation, James Island 51
McLeod Plantation History 51
Objectives of the Grand Experiment: People and Science Based 53
Results of People-Based Experiment 56
Results of Science-Based Experiment 57

**11 Third Site of Large-Scale Sweetgrass Plantings: Dill Sanctuary,
       James Island**   63
       Dill Sanctuary History   63
       Results of Dill Plantation Field Experiment   64
       Reevaluation of Large-Scale Sweetgrass Plantations   67

12 Alternative Ways to Access Sweetgrass...   69
       Dependence on Private Beachfront Communities to Allow
       Access to Sweetgrass   69
       Municipal/Commercial Plantings   69
       Restoring Natural Habitats Along SC Beaches   72
       Army Corps of Engineers Dune Vegetation Shore Protection Project.....   72
       Sweetgrass Planting Logistics at Folly Beach, SC   75
       Folly Beach Sweetgrass Planting Locations   76
       Folly Beach Sweetgrass Development Over the Years   77
       Grand Strand Sweetgrass Project..   81
       Grand Strand Sweetgrass Planting Locations   81
       The Happy Ending to a Long Struggle   82

13 Sweetgrass Biology..   83
       Sweetgrass Botany: What's in a Name?   83

14 Sweetgrass Horticulture: Environmental Considerations.................   87
       Distribution in the USA and Habitats   87
       Plant Longevity   87
       Soils...   88
       Heat/Drought Tolerance   91
       Pest Pressures   91
       Fertility Responses...   92

15 Seedling Cultural Practices..   95
       Clump Separation to Produce Sweetgrass Plugs   95
       Seed Germination and Seedling Culture..   97
       When to Seed   98
       Containers   98
       Seeding the Media...   98
       Seedling Fertilization   101
       Seedling Watering...   102
       Seedling Hardening...   102
       What Is a "Quality" Sweetgrass Transplant?..   103
       What Are the Problems in Growing Sweetgrass Transplants?   104
            Fertility..   104
            Low Light..   104
       Seeding Only a Few Seeds per Container...   104
       Overseeding   104
       Greenhouse Growth Period Was Too Short   105
       Excessively Long Leaf Production   105

16 Field Production Practices 107
Timing of Spring Planting in the Field 107
Soil Conditions.. 107
Fertilizer Additions in the Field.. 108
Watering 108
"Benign Neglect" 108
Weed Control 109
Light, Shade, and Competition 110
Plant Spacing in Fields Planted for Basketmakers' Use 110
Timing of Fall Planting in the Field.. 111
Growth Practices in the First Year After Planting 111
Growth Practices in the Second Year After Planting and Beyond........... 112
Changing Sweetgrass into "Frankengrass" 112
Harvesting 113
Yearly Renovation.. 113
Longevity and Decline of Sweetgrass Plants in Cultivation.................... 114

17 Concluding Thoughts... 115
What Mysteries Are Left to Unravel?... 115

18 Afterthoughts... 117

References 119

Chapter 1
Introduction

Charleston, South Carolina, is considered to be the "Jewel of the South" by vacationers and, of course, its lucky inhabitants. The city was founded in 1670 and the Preservation Society registers more than 2,000 structures in the old downtown as historic landmarks.

The "Holy City" (Fig. 1.1) contains hundreds of churches and has always encouraged religious freedom. This enchanting city has many lovely old homes and gardens throughout its old and winding streets making Charleston an absolute pleasure to explore study and admire. Even older than most homes in Charleston is the ancient tradition of sweetgrass (*Muhlenbergia sericea* (Michaux) P.M. Peterson) basketmaking, yet basketmaking is truly the only "live" viable symbol or relic of the extinct plantation days of colonial times. Originally these graceful baskets provided useful, practical objects or "work baskets" for agricultural and household uses on the plantation; today, they have evolved as souvenirs treasured by tourists and elegant objets d'art. At the bustling Charleston City Market frequented by visitors and vacationers from around the world, basketmakers are a standard presence selling their baskets and the curious visitor becomes acquainted with the ancient art of African coiled basketry (Fig. 1.2). The diversity of shapes and forms of these wonderful baskets are extensive with new forms constantly being "invented" (Fig. 1.3).

The future of this wonderful folk art is in jeopardy. Sweetgrass once was readily available and abundant in the Charleston area, but today, sweetgrass is very scarce and more difficult to find. This book recounts the story of a 20-year struggle to learn the mysteries of how to domesticate this wild plant and how to help rectify the scarcity.

Colonial Charleston was a very prosperous city built on many agricultural industries beginning with rice in the late 1600s and early 1700s, then indigo, and lastly cotton. It was determined in colonial times that the exotic crop rice could be grown with great success in the Southeast, especially the Charleston area.

With its indigenous warm climate, a long growing season, coupled with extensive wetlands, the area was a natural for growing the lucrative rice crop (Fig. 1.4). Low Country plantation owners of European descent knew little of the agronomy of

Robert J. Dufault, *Stalking the Wild Sweetgrass*, SpringerBriefs in Plant Science, DOI 10.1007/978-1-4614-5903-3_1, © Robert J. Dufault 2013

Fig. 1.1 Overview of the peninsular city of Charleston South Carolina (Source: http://southern-boating.com/blog/wp-content/uploads//2011/05/Charleston-Aerial-View_Charleston_Area-CVB_14MB.jpg)

Fig. 1.2 A typical basketmaker at the City's Market selling a wide variety of sweetgrass baskets (Source:http://travellogs.us/2010%20Logs/South%20Carolina%202010/10-61%20Charleson%20SC/10-61%20Charleston%20South%20Carolina.htm)

Fig. 1.3 The diversity of shapes, forms, and styles of sweetgrass baskets is tremendous as basket-makers have "invented" new ways to express their art

growing this unusual crop and, of course, did not have the work force to produce it anyway. However, their lust for the money and power that rice would bring spurred the need to achieve the goal of growing this crop. Enter the slavers who knew the existence of the great Rice Kingdom in western Africa. Only certain African people, however, especially those from Senegal, were brought to South Carolina as early as 1505 with the advent of the slave trade in the New World (Fig. 1.5). Although the variability of African baskets is as diverse as the over 800 different ethnic groups that compromise African people (Vlach 1978), the Africans from the rice coast sewed unique baskets tailored for rice processing, storage, and other utilitarian purposes. Today, we still see the same type of baskets being sold by the descendants of the enslaved Africans of centuries ago. By the end of slavery around 1808, Africans were abducted from areas extending down to Central Africa and Mozambique. Rice was a staple crop in Africa and Africans had cultivated rice from time immemorial and perfected its production. Slavers identified these people, the ancestors of today's

Fig. 1.4 Historic artist rendering of enslaved Africans tilling rice fields in the Low Country of South Carolina (Source: http://www.elowcountry.com/blog/index.php/2012/01/30/slavery-and-rice-on-the-santee/)

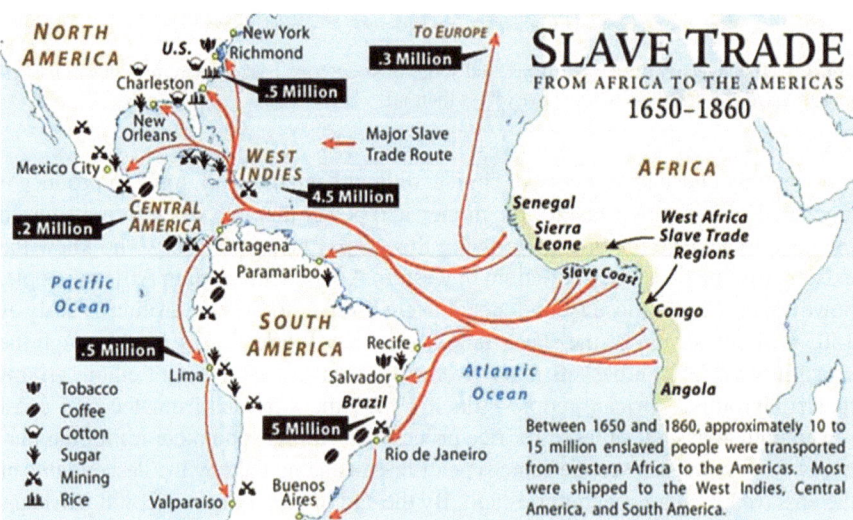

Fig. 1.5 Slave trade from Africa to the Americas from 1650 to 1860 (Source: http://www.slaverysite.com/slave%20trade.htm)

Low Country basketmakers, and purposely abducted them for sale to plantation owners for use on rice plantations in the New World. Slavers were paid premium price for Africans from the African Rice Kingdom of the Windward coast (Senegal to the Ivory Coast) to the mouth of the Congo River (Gabon, Zaire, and Angola) (Littlefield 1981).

A man or woman who made baskets was worth more than one who did not, age, strength, and other skills being equal (Charleston Gazette and Advertiser 1791). Since colonial South Carolinians knew little of rice culture, the success of the American Rice Kingdom is credited to the enslaved Africans (Joyner 1984). Charleston's early rice-plantation owners desperately needed the labor of those who did know of rice culture. With the enslaved African people, came the expertise to produce rice from the planting, harvesting, and processing and knowledge and talents for hunting, agriculture, and art. The craft of African coiled basketry crossed the Atlantic in slave ships and took root in the new land, but this action, of course, had devastating consequences on the peoples living in the rice-growing regions of the West African Atlantic coast. Though many linguistic, religious, and cultural bonds were destroyed by the slave trade, enslaved Africans didn't abandon all their ethnic traditions once they reached America's shores.

The first African-American basketmakers sewed wide circular trays, called fanners (above), to separate individual grains of rice from chaff, or husks (Fig. 1.6). After the grain was threshed and pounded, it was thrown into the air in the fanner so the wind would blow away its rough outer shell (Fig. 1.7). Other types of baskets were used for foodstuffs. In fact, the baskets made by the slave communities were so necessary for agricultural production that bartering and selling them among different plantations provided Charleston's planters with an extra source of income. African coiled basketry was critical in the processing of rice. All through the antebellum period, both men and women made baskets. Men were responsible for crafting containers and fanner baskets made of sturdy bulrush grass, while women constructed household baskets using the lighter, more flexible sweetgrass. Although men still learned basketry in the postemancipation era, fewer felt compelled (or had the time) to practice it so basketmaking was mainly left to women.

The roots of the beautiful art of Low Country sweetgrass baskets were borne out of the holocaust of slavery in the colonial South. In most of the Mount Pleasant families in which the basketmaking tradition survived, family members carried the skill from generation to generation. Carefully passed down from mother to daughter or grandmother to granddaughter, basketry connected these African-American families to the African continent and a heritage that for other African-Americans seemed far more distant. Sweetgrass basketmaking is one of the earliest traditional crafts with a rich documented history from "carryover" from African enslavement and plantation days to the present (McKissick Museum 1988). The art, craft, and significance and legacy of producing these historic baskets has been treasured, beloved birthright inherited by generations of basketmakers, the direct descendants of enslaved Africans of plantation days. To value the significance of this craft, it is important to understand and appreciate the spiritual fervor and passion that basketry holds for the individual and the community. After all, this is far more than just an

Fig. 1.6 Simple fanner basket used to separate rice chaff from seed

Fig. 1.7 Actual use of fanner basket in cleaning rice seed postharvest

issue of making baskets for economic gain in the Basketmaking Community but preserving a "living link" to their African heritage. Even today this cultural bond remains strong. Though the raw materials used in African and African-American coiled basketries differ, their construction remains remarkably similar. African-American basketmakers have traveled to Senegal and returned with stories of having seen almost identical methods used by their West African counterparts; today the technologies survive on this continent mainly around Mount Pleasant, in the heart of the American colonial rice-growing region. Although the materials used are different in the USA, the form and function of the African counterparts of sweetgrass baskets are unchanged (Mary Jackson, personal communication).

After the Emancipation Proclamation in 1863, former slaves were determined and dedicated to preserve the African traditions of their ancestors. To the basketmakers, sweetgrass basketry symbolizes (1) their heritage and connection to their African ancestry; (2) an arduous and proud history and remembrance of people who survived centuries of oppression; and (3) a source of independence, income and pride of producing wonderful objects of art so treasured by so many. They have made a reaffirmation of these goals with each new generation of basketmakers with sweetgrass basketry representing the heart and keepsake of their devotion.

Sweetgrass, the main structural material used in African coiled basketry, is a native, perennial, warm-season grasses found growing sparsely along the ocean in narrow bands between coastal sand dunes from North Carolina to Texas. With the help of relatives, basketmakers have, for 300 years, gathered the green sweetgrass leaves in bunches from familiar and nearby coastal dune areas adjacent to the ocean front in Mount Pleasant and the Low Country of South Carolina. The leaves are allowed to dry out for a week to 10 days. The slender sweetgrass leaf blades are sown into successive coils, not woven, row upon row, with strips of leaf materials from the heart of new growth of the palmetto tree, the state tree of South Carolina. Sometimes they lace coils of sweetgrass with pine needles for color and contrast. They draw their inspiration from everyday objects like cooking vessels, dishes, and flower vases, and their creations come in a startling variety of shapes and sizes that have won national acclaim.

Rosengarten (1987) has provided an excellent review of basketry techniques. Coiled basketry involves sewing and stitching unlike other types of baskets that are woven. Specifically, each basket begins with a small knot of long-leaf pine needles (*Pinus palustris* Mill.) or the fine-threaded sweetgrass. Pine needles may be used throughout the basket and provide a russet color that contrasts well with the more golden/yellow sweetgrass. Coarse, thicker-gauge black rush (*Juncus roemerianus* L.), also known locally as "bulrush, rushel, or needlegrass," may be added to the inside of the baskets for strength. Black rush turns a rich tawny color when dried. Strips of palm leaf (*Sabal palmetto* Lodd.) are used to sew the rows of coils to each other. A hole for the palm strips is made with a bone (a spoon with the bowl removed and the end smoothed and polished), nail, or bagging needle (Fig. 1.8). In olden times, a carved animal bone may have been honed to do this task, hence the name "bone."

Shape of the baskets is created by building upon the foundations, one row at a time and then row upon row. The coils of material must be constantly fed with new grasses to maintain a constant foundation of uniform thickness, all the while the symmetry of the coiled palmetto being adjusted to produce a "sun burst" effect as much as possible. The strength of the basket depends on how firmly the stitches are pulled. Today, baskets of great variability are constructed by modifying the amounts of these materials to create subtle changes of tone and radical change of texture.

Fig. 1.8 Basketmaker using a metal "bone" to make an opening to sew a strip of palmetto threw a sweetgrass coil to join row upon row (Source: http://ghettogardens.blogspot.com/2010/03/charleston-made-goods.html)

The explosion of growth in the Charleston area by transplanted tourists and industries has caused a boom in urban, and beachfront development around Mt. Pleasant, SC, the ancestral "home" of sweetgrass basketry in the " new world," has destroyed much of the natural habitats of sweetgrass. Other traditional gathering places in the barrier islands off the coast of Charleston have been developed as beach front resorts or private communities with restricted access. Much of the new housing along the ocean is built in the same location as native sweetgrass fields. Basketmakers now have to travel hundreds of miles to Georgia or Florida to find adequate supplies or buy harvested sweetgrass from members of the community that still know hidden places where sweetgrass still grows. Many basketmakers are old and these long, expensive, arduous trips are impossible and frustrating. This frustration has fueled the loss of basketmakers.

The Mount Pleasant Basketmaking Community has become known throughout the USA and abroad for their art; sweetgrass baskets are a symbol of Coastal South Carolina. Postcards and photographic illustrations commonly depict basketmakers and their art. African coiled basketry is objects of scholarly research and commonly is featured in newspaper and magazine articles in the USA and abroad. Sweetgrass baskets are displayed in art galleries and traveling exhibits and have been displayed in the Smithsonian, Vatican, and Gibbes Art Museums in Charleston and other museums.

Basketmakers have always wondered if they could cultivate sweetgrass inland similarly to garden plants, but individual attempts to transplant the plant were not

Fig. 1.9 A basketmaker at the Charleston Market working diligently on a basket surrounded by a variety of her creations

successful. By 1988, the plight of the basketmakers was common knowledge due to timely newspaper articles. In 1986, I moved to Charleston and as any new resident of the City does, I visited the City Market downtown between East Bay and Meeting Streets. The Market is a bustling tourist stop with an extravaganza of all kinds of merchandise including sweetgrass baskets. There, I saw women making baskets of great intricacy and beauty. Alongside of these women, I could see their raw materials, clumps of long broom-like grasses (sweetgrass and bulrush) and bundles of what looked like whips, which were the palmetto leaf strips used to "sew" the rows of grasses together and bags of copper-colored pine needles.

What the women were doing, to the untrained eye, was not basket weaving but more appropriately termed basket sewing (Fig. 1.9). So I was mesmerized by watching these "weeds" being sewn together into fine objects of artistic design and function. I wanted so much to talk to the women about their work but in many cases, their eyes would be riveted to the task of sewing these baskets and I was too timid to ask "stupid" questions. Little did I know that within a few years, my personal art of horticulture would be called on to help these artists in the preservation of their craft!

I have been involved in horticultural research at universities on vegetable crops since 1978. In my over 32 years of experience, I have worked on the production of many crops ranging from asparagus to zucchini, agronomic, medicinal, and floricultural crops. In 1986, I assumed a 100% research position with Clemson University at their Coastal Research and Education Center (CREC) in Charleston, South Carolina. In 1989, I became involved in learning more about the horticulture of the native sweetgrass species and over the years, learned how to grow it as a "domesticated crop." Many of our vegetable crops and grains originally started out as wild plants that eventually were "tamed," improved sometimes through breeding programs, released by breeders as cultivated varieties, and grown in cultivated rows on farms. Why not domesticate sweetgrass using old nonbreeding approaches?

The Mount Pleasant group's determination to protect its sweetgrass from extinction excited me as a horticulturist, and the group's enthusiasm alone might well have been enough to attract me to this conservation project. Very early in my process, I recognized that assisting the basketmakers would further my own scientific research, but learning the cultural value of what was at stake gave me an even greater desire to become involved. The basketmakers and I decided later in the process to organize experimental large-scale sweetgrass plantings. We faced many difficulties. The first challenge was to figure out how to grow it. On coastal beaches it grows on pure sand in a desert-like situation with very little water, rooting itself deeply and sending up leaves that are highly resistant to heat and salt and it needs little fresh water. Since it lives near the shoreline, the plant has few natural enemies or competitors except land developers. These attributes make sweetgrass exceedingly resilient—in its natural habitat. Interesting, the tenacity of wild sweetgrass is very similar to the tenacity of the descendants of enslaved Africans who ardently wished to perpetuate their craft and the plant critical in its construction.

My objective from the beginning of my involvement with sweetgrass was to learn how to grow this wild plant as a row crop. Over two decades have elapsed and like fine wine, knowledge and experience have "aged" and much knowledge has been discovered about the mysteries of this once illusive wild dune plant inhabiting the Southeast coast. Just like the basketmakers that sew baskets, one row upon another row, the secrets of sweetgrass culture have been added "one row upon another row." Over the years, my knowledge of the "how to's" and the "not to's" have grown greatly.

This book records the 20-year struggle to learn how to domesticate this wild dune plant through cultivation and farming including efforts to replenish the original wild habitats as another way of increasing the supply and reducing scarcity. This book reviews the difficulty of trying to involve a group of artists to learn and teach the mysteries of the plant and take responsibility for its cultivation. It is a book of trials and errors, failures and triumphs to finally learn enough to begin rectification of a decades-old problem of an ancient folk art's survival into the twenty-first century. The first half of the book recounts the historical voyage the work and research took and the second half of the book explains the horticulture of how to grow the plant and its cultural needs for quality production as well as cultural aspects that spell doom for this plant if not attended to. I hope you enjoy this story and that my story may act as a template and guide to follow for anyone else attempting to prevent a critically important economical wild plant used in historically significant folk art—what to do to, what to avoid, and what to expect when a plant's future and the folk art depend on aggressive human intervention and science.

Now in my retirement years, I feel that the two decades of my work, observations and history of working with sweetgrass, need to be catalogued and preserved. The art of making African coiled baskets extends over three hundred years in the USA and time immemorial in Africa.

Yet, the work I have done stalking the wild sweetgrass and domesticating this plant needs to be preserved, for it is part of this tradition; serves as a starting point for other scientists or "helpers" that follow; and is my contribution to the Sweetgrass Community who still produce, as their forebears, such wonderful objects of art and utility.

After two decades, there is so much now known about this plant and yet there still is much to be learned. Sometimes I still get telephone calls about sweetgrass and spend hours educating interested folks on what has been tried, what works, what doesn't work, and the obstacles that still loom in the future. I feel a personal responsibility now to get all this horticultural and background information, history, facts, and fantasies down on paper.

If genuine, serious service can be given to the Sweetgrass Basketmaking Community, I have a responsibility to share what I have attempted and know and to save those people in the future their time and frustrations to not repeat the same mistakes I made. So in the following chapters, I will take you on a discovery of the process I have traveled in my quest to stalk and domesticate this wild plant and to help alleviate the sweetgrass shortage problem that has plagued this community for 4 decades or more.

As I have said to many people who still remember my earlier work on growing sweetgrass at McLeod and Dill Plantations, I never gave up on my task of helping solve this multifaceted problem of supply, and yet even in the writing of this book, I feel my story does have a "happy" ending. Documenting here all the techniques and knowledge that have stood "the test of time" and can be depended on to "work" is an additional form of success that I want to share with the Basketmaking Community and all those who will follow in the future and who may continue the work with sweetgrass.

Chapter 2
The Beginnings of Change

There were many events that led up to the flurry in sweetgrass work that began in the late 1980s before my entrance into the arena (Fig. 2.1). Of course, the first event that contributed to the stress was the explosion of coastal development in the late 1970s and early 1980s which even continues today.

Wild areas along the coast, once plentiful with traditional sweetgrass-harvesting habitats, were gobbled up for the construction of million dollar beach houses behind gated entrances (Fig. 2.2). In many of these cases, this new housing along the ocean was placed in the same location as natural sweetgrass habitats, and these fields were "plowed down" (Fig. 2.3).

Sweetgrass Conference, 1988

The beginning of real change to initiate some action and understand the issues facing the basketmakers did not commence until the Sweetgrass Basket Conference which was held at the Charleston Museum on March 26, 1988 (Fig. 2.4). The conference was organized by the South Carolina Folk Arts Program at McKissick Museum, University of South Carolina, with the cooperation of the Steering Committee of Sweetgrass Basketmakers, Avery Research Center for Afro-American History and Culture, College of Charleston, National Trust for Historic Preservation, and Seabrook Natural History Group. The conference was funded in part by grants from the National Endowment for the Arts, Folk Arts Program and the Ruth Mott Foundation. This conference evaluated the impact of public policy and development on Charleston County basketmakers' access to sweetgrass resources (Fig. 2.5). In addition, the conference addressed many issues facing the basketmakers with the interaction of officials, environmentalists, basketmakers, biologists, and the public.

This unprecedented conference was long in coming. Some of the earliest events that led up to this conference began in the early 1970s when two young anthropologists, Greg Day and Kate Young, spent a significant time in the Basketmaking

Robert J. Dufault, *Stalking the Wild Sweetgrass*, SpringerBriefs in Plant Science, DOI 10.1007/978-1-4614-5903-3_2, © Robert J. Dufault 2013

Fig. 2.1 Over the years, many newspaper and magazine articles have been published describing the dwindling sweetgrass supply as a victim of ocean from developers

Fig. 2.2 Sorry, property owners only! Gated communities still contain some sweetgrass habitats that once were accessible and free to harvest

Fig. 2.3 A native sweetgrass field on Kiawah Island, South Carolina, was bulldozed to make way for beachfront house. See the remnants of the sweetgrass field on the left side of clearing

Community talking with and studying the art of basketmaking. One of their tasks was to assemble a collection of sweetgrass baskets for the Smithsonian Institute. As a result, examples of sweetgrass baskets are now in museums and this act elevated the general public's awareness, acceptance, and appreciation of the basketmaking art. Later, attention and interest in the basketmaking art form continued to grow, and in 1984, McKissick Museum hired Dr. Dale Rosengarten to study the history and practice of Low Country basketry. As a result, this led to the development of "Row Upon Row: Sea Grass Baskets of the South Carolina Lowcountry" exhibit organized by the University of South Carolina's McKissick Museum in Columbia (Fig. 2.6). Ultimately, this exhibit was transformed into a book by the same name by Dr. Dale Rosengarten. This seminal work excellently described the origin and evolution, folk life, techniques, and the state of the art of Low Country African-coiled basketry.

The idea for the conference came about in October 1986 at a meeting of folklorists, including Dr. John Vlach from George Washington University, Mrs. Bess Lomax Hawes of the National Endowment for the Arts, and Dale Rosengarten[2]. This conference was the first organized effort to find answers to deal with the regional problems affecting this ancient folk art. In many cases, developers, conservationists, and managers of public lands and wildlife preserves did not recognize sweetgrass as a scarce and significant resource that may exist on their lands. Basketmakers were very assertive and verbal, expressing their frustrations with the many pressures threatening the future of this craft.

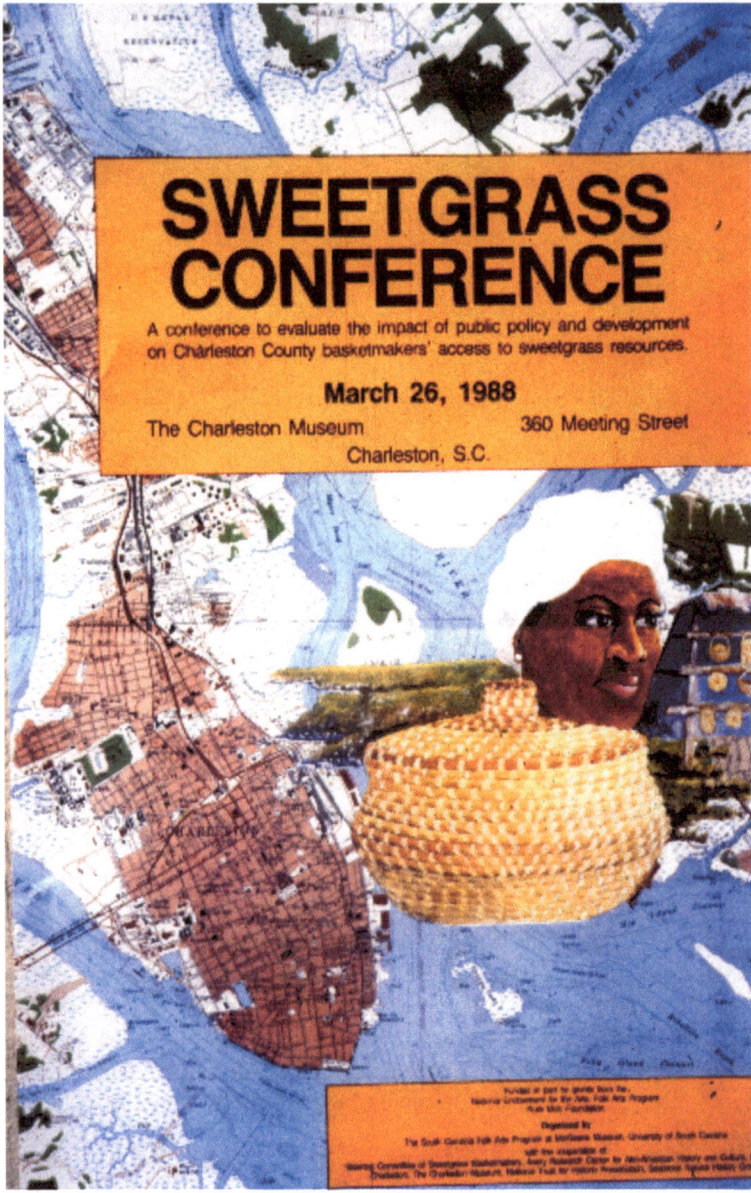

Fig. 2.4 The goals of the sweetgrass conference were to draw attention to a culturally and economically vital plant and to encourage land management strategies which favor the grass and make it accessible to basketmakers and their families

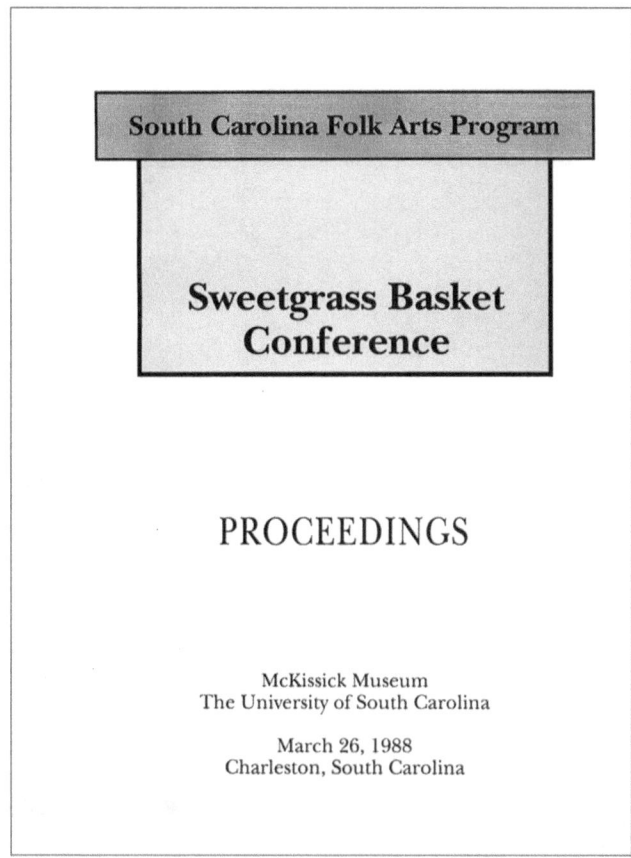

Fig. 2.5 Proceedings of the historic conference

The Findings of the Conference

The following briefly summarizes some of the findings revealed during the conference:

Sweetgrass is not an endangered species but it does seem to exist in limited quantities along the Southeastern coast. Biologists stated that they could not, at that time, state how the plant propagates, whether it can be transplanted, and under what conditions it thrives. Recommendations from biologists included the following:

(a) Conduct a biological assessment of the plant.
(b) Develop a map of sweetgrass populations.
(c) Determine the threat to the population.
(d) Identify the ownership of the land on which sweetgrass grows.
(e) Identify appropriate locations for transplantation.

Fig. 2.6 Dr. Dale Rosengarten's rich publication is an in-depth study of her 18 month research project on this folk art form which includes both research and history of this flourishing African-American craft and a survey of its current status

Some of the basketmakers indicated that they had tried transplanting but with little success. It was also concluded that there still was quite a bit of sweetgrass growing in areas near Charleston, but the gatherers cannot get to these habitats. The developers have captured the coast. Areas where people have gathered sweetgrass for hundreds of years are now off-limits. Developments have, in effect, made what was once public coastal space a private domain.

Basket stands along Hwy 17, the old Ocean Highway from Florida to New England, are being forced from their traditional highway locations by development. Basketmakers have been selling their baskets on the shoulders of Highway 17 in Mt. Pleasant since the 1930s. The basket stands are part of the visual and cultural landscape of Mt. Pleasant, yet their placement may be hazardous to basketmakers and their continued presence may be growing extinct with development in Mt. Pleasant.

This land is under the jurisdiction of the Highway Department and not the owners of the land behind it and not the town of Mt. Pleasant. The future existence of these stands as development explodes needs to be protected as part of the cultural gift they provide and as a means for basketmakers to sell their products to passing tourists.

The Steering Committee, which was composed of basketmakers, decided to continue their work after the conference and formed the Mt. Pleasant Basketmakers' Association on April 5, 1988. The main purpose of the organization was to promote, protect, and preserve the traditional African art of sweetgrass basketmaking. After the conference, many promises were made to continue the work to protect roadside stands, attempt to cultivate sweetgrass in various locations, and survey existing native stands of sweetgrass. Moreover, this conference was the culmination of making people aware of a problem that they didn't know existed.

Chapter 3
My Time to Get Involved

The Honorable Lucille Whipper, District 109, SC House of Representatives, contacted Clemson University main campus by mail in February 1988 and sent a copy of the conference agenda. Unfortunately, I was not informed of this conference. After the conference, however, Keith Salvo, the Plant Materials Specialist for the Soil Conservation Service (SCS), developed the "Sweetgrass Conservation Field Trial Proposal." In February 1989, the SCS developed the "Economic Development Resource Conservation and Development Measure Plan" for potential funding through the Low Country Resource Conservation and Development Council. Ann Christie, District Conservationist with the SCS, contacted Bill Yates, Clemson University Extension Program Coordinator, and asked if the Coastal Research and Education Center in West Ashley, south of Mt. Pleasant, could be the site of experimental plantings. In February 1989, I was asked if I would be interested in studying sweetgrass propagation and culture. Based on my earliest exposure to observing the basketmakers in the City Market and being enthralled with what they were doing, I immediately jumped on the idea! There began my journey with sweetgrass that lasted for 21 years.

On February 3, 1989, a meeting was held to prepare and discuss the measure plan mentioned above, and a variety of professionals, including myself, were invited from many different agencies to include:

Chris Brooks, Coastal Council
Mel Goodwin and Sandy Goodwin, Sea Grant Consortium
Keith Salvo, Steve Edwards, and Ann Christie (chairperson), SCS
Dr. Richard Porcher, the Citadel Biology Dept.
Linzie Muldrow, 1890 Extension at SC State College
Dale Rosengarten, basketmaker researcher
Mary Jackson, president, Mt. Pleasant Basketmakers' Association
Chip Boling, Clemson Extension Service

Robert J. Dufault, *Stalking the Wild Sweetgrass*, SpringerBriefs in Plant Science, DOI 10.1007/978-1-4614-5903-3_3, © Robert J. Dufault 2013

Similar to the conference, many promises were made to continue the work to:

1. Research the cultivation of sweetgrass in various locations
2. Survey existing native stands of sweetgrass
3. Develop needs assessments (who needs it, how much, and where and when)
4. Produce digitized coastal maps to delineate existing sweetgrass fields
5. Inventory existing sweetgrass populations by exploration
6. Establish permanently located roadside markets to sell baskets

Change takes time and some attempts were made to satisfy those goals with mixed levels of success. In some cases, even with all good intents and purposes, the promises were not fully kept. Over time, the original group became fragmented and people retired or transferred to other jobs; however, a core group of "diehards" remained for a few years working to solve problems, and I was one of these. Over the years, other people came along and "discovered" the sweetgrass plight, engaged for a couple years, made some contributions to our knowledge base, and moved on with other projects. At least for sweetgrass cultivation and solving the supply issue, by 2010, there is not a single "answer" to this predicament, but we now have an array of attractive options that did not exist in 1989.

What Were the Sweetgrass Basketmakers' Community Needs?

One of the few individuals that carried on after the conference was Linzie Muldrow, SC State College. He completed an assessment of the Sweetgrass Basket Community in Mt. Pleasant. In July 1990, Dr. James Walker and Linzie Muldrow offered the following information about the Sweetgrass Basketmakers Community (based on that time frame and, of course, circumstances have now changed at the writing of this book in 2010):

– Approximately 300 families were involved in the basketmaking tradition in the late 1980s with approximately 75% full time and 25% part time.
– Estimated that about 15–20 acres of sweetgrass are needed to sufficiently meet the needs of the Sweetgrass Community.
– Basketmakers travel approximately 500–700 miles to secure sweetgrass with length of trips from 3 to 4 days at a cost of $200 to $300 per trip.
– Hurricane Hugo in September 1989 severely impacted their palmetto supply, destroyed many roadside stands, and washed away existing sweetgrass habitats near Mt. Pleasant (Fig. 3.1).
– Basketmakers desire more public awareness of their problems and greater support from local government to promote sweetgrass basketry.
– Serious reservations exist about local government that support preservation of sweetgrass basketmaking yet, at the same time, support development interests in the area, especially roadside stands.
– Concerns about the use of the word "*sweetgrass*" which is used indiscriminately by developments or entrepreneurs, linking their projects to the ancient tradition without any real connection to it or input from the Basketmaking Community.

Fig. 3.1 Sweetgrass basket stands are numerous on Highway 17 Mount Pleasant, SC

- Basketmakers want to be involved in the economic development plans of their community. Any plans that will affect their tradition, they request representation.
- Desire that a buffer zone be established along Hwy 17 to protect the spaces where basketmakers have traditionally sold baskets in stands.
- Desire to purchase land to establish a sweetgrass basket market along Hwy 17.
- Desire cultivation research to be conducted on their land or their backyards.
- Desire to find sweetgrass fields within SC and to have open access to them.
- Expressed concerns about the younger generation not showing much interest in learning the basketmaking craft.
- Concerned with developing a strategic plan for expansion and marketing of baskets in the Mt. Pleasant area.
- Develop a brochure to educate the public/local businesses/developers on the importance of basketmaking in their community.
- Accessing local, state, and federal funds to support the sweetgrass tradition and organizations.
- Have soil tests done on basketmakers' land to determine the feasibility of growing it there.
- Want control of who grows sweetgrass in their area.
- Construct and display signs throughout Mt. Pleasant depicting the area as home of sweetgrass basketry.

Chapter 4
Getting to the Grass Roots of Cultivation

In my estimation, one of the most significant early attempts to help the basketmakers with their supply issues was through the work of Keith Salvo, SCS. In fact, after being involved in the sweetgrass work from 1988 to his retirement from SCS in 1991, he struggled for a few years to set some groundwork and to be the first person to attempt the domestication of sweetgrass. Years later, he was nominated by the Mt. Pleasant Basketmakers' Association and the SC Folk Arts Program and received the USDA "Unsung Hero" award. I agree with their choice. Although in the long run, Keith's work at establishing a large plantation of sweetgrass failed at Palm Key, with all failures, comes positive knowledge and much was learned through that first endeavor.

Keith wrote and initiated the "Sweetgrass Conservation Field Trials." This was to be a 5-year study with the objectives to:

- Determine cultivation and technical information necessary to assist the basketmakers to establish and manage this natural resource for production purposes in their local area.
- Determine conservation and other uses of sweetgrass.

Keith's first step in this process required traveling to areas where sweetgrass fields grow naturally on the coast. In March 1989, Keith made collections of wild plants from Sapelo Island, GA; Kiawah and Seabrook Islands, SC; and McClellanville and Georgetown, SC, and samples were sent to the Plant Materials Centers in Brookings, FL, and in Americus, GA. In September 1989, more plants were collected at Sea Pines, SC, and 33 plants were transplanted in small plots at the Clemson University Coastal Research and Education Center, Charleston, SC (Fig. 4.1). For the following years, I was in charge of growing these plantings, giving water, cultivation, and fertilization using the same cultural practices you normally would give a vegetable crop.

Robert J. Dufault, *Stalking the Wild Sweetgrass*, SpringerBriefs in Plant Science, DOI 10.1007/978-1-4614-5903-3_4, © Robert J. Dufault 2013

Fig. 4.1 In September 1989, Keith Salvo, Plant Materials Specialist with the Soil Conservation Service, planted sweetgrass ecotypes collected from Sea Pines, SC, at the Clemson University Coastal Research and Education Center, Charleston, SC

First Site of Large-Scale Sweetgrass Plantings: Palm Key Resort, Ridgeland, SC

In October 1990, Keith developed a new project to establish sweetgrass at Palm Key Resort near Ridgeland, SC. This location was hoped to be a valuable local supply of sweetgrass for continuation of the craft. Palm Key Resort is a private 350 acre island that hugs the Broad River as it reaches toward the sea. This is a pristine resort dedicated to a "nature experience" for guests among its woodlands, freshwater lakes, waterways, and marshes. This location would be the first time a large field of sweetgrass would be planted into cultivation. A one acre tract of land was prepared for planting sweetgrass. The needed plants were dug from Kiawah Island, SC. The Mt. Pleasant Basketmakers' Association was involved in digging sweetgrass and planting it at Palm Key on November 16–17, 1990. The sponsors for this event included the Beaufort–Jasper County Water Authority, Jasper Soil and Water Conservation District, Clemson University Cooperative Extension Service, Soil Conservation Service, and Low Country Resource Conservation and Development Council. By July 1991, however, the planting at Palm Key failed due to lack of proper maintenance and poor soils. Being so distant from the basketmakers and others who could have monitored the field was a grave disadvantage. More seriously, however, a soil test of the Palm Key field indicated a very acidic pH of 5.1 and a very high manganese level. A low pH unfortunately makes manganese, iron, boron, copper, and zinc more available in the soil, and a high level of these elements can cause toxicity and actually kill the plants. A 5.1 pH can also cause some other essential elements in the soil to become "overabundant" such as manganese to cause toxicity. A low pH may cause some elements such as nitrogen, phosphorus, potassium sulfur, calcium magnesium, and molybdenum to become deficient or even "unavailable," and the plants could "starve" for those essential elements. The simple act of liming the soil before planting may have helped the sweetgrass grow by raising the pH to 6.5 which for most horticultural crops is ideal. Liming was not performed at Palm Key though. In sadness, that attempt was abandoned, but some important lessons were learned at Palm Key even though the planting died.

1. Specifically, sweetgrass cannot tolerate acid soils. This was unknown before Palm Key. Therefore, it is very important to gather soil samples from natural habitats on the dunes of high-quality sweetgrass to act as a "model" of what soils should contain before choosing a site for cultivation.
2. Soil tests must be taken before planting and pH adjusted to the range that sweetgrass thrives in. Potentially, soils may be amended with synthetic nutrients to artificially mimic the soil fertility of natural sweetgrass habitats.
3. Sweetgrass needs to be monitored and tended after planting. Sweetgrass fields cannot be planted and walked away from except for plantings made at the beach which will be discussed later.
4. Sweetgrass cultivation must be planted very close to the Mt. Pleasant Basketmaking Community so they can be involved in monitoring the crop's progress.

The CREC planting made by Keith continued to flourish after a year and the future looked bright. Keith's sweetgrass project was to have 5-year duration until 1994 (Fig. 4.1). There were aspects of the multiple plantings he was to evaluate, such as fertility requirements, disease and insect problems, and harvesting schedules of these plantings under cultivation. But Keith decided to take an early retirement in 1992, so we lost our prime horticulturist in charge and the Sweetgrass Conservation Field Trial had an abrupt end. SCS did not replace Keith in this work, so at that point, I had taken such a great interest in this crop; I continued tending these plants, "beefing" them up, and they eventually became very luxurious.

Chapter 5
Getting More Involved

Since my area of expertise is vegetable crop culture, I applied the same array of production strategies I would use for any high-value vegetable crop on sweetgrass. The sweetgrass I inherited from Keith received copious nitrogen fertilizations throughout 1989 and 1990 and three applications in 1991. The resulting plants were gloriously huge and looked wonderful to the eye. Little did I know back then, that was NOT what the sweetgrass needed for high-quality fiber strength for the basket-making tradition. In hindsight, the way sweetgrass will be used eventually dictates how it should be grown. If it is to be used as an ornamental, copious fertilization is appropriate. If it is to be used for sweetgrass basketry, fertilizer absence after the first year is a cardinal rule of culture.

After 21 months of producing luxurious growth at CREC, I contacted Mary Jackson, president of the Mt. Pleasant Basketmakers' Association, to come visit the planting on June 7, 1991. I remember vividly, driving Mary down the dirt road to the sweetgrass plot, and within about 300 feet, it became visible. Mary became very excited about how great it looked from a distance. And it did, but to my great dismay, when Mary tried to harvest this grass, it all broke apart in her hands. The sweetgrass became so succulent and brittle, low in fiber, lush, and very fragile for pulling. The leaves unfurled, and rather than being very tight, wirelike "threads," they opened up like lawn grass into flat blades. I remember well; Mary said *"You changed it...this isn't sweetgrass!"*

And indeed, I did change it but it was genetically sweetgrass, but I learned the first great lesson of growing sweetgrass and that was—after the first year—withdraw any fertilization and use the cultural technique I have termed "benign neglect." Benign neglect simply means do not fertilize and allow the plants to be stressed (more discussion in File Production section).

It is important, however, beginning the first year after planting, to keep an eye on the plants, treat for fire ants, weed them, clean the plants each year in winter of dead leaves, and water them in extreme drought (but normally let them "get watered" by natural rainfall). One of the greatest frustrations with collecting sweetgrass from plants used for ornament in the landscape (gold courses, parks, residential areas, etc.)

Robert J. Dufault, *Stalking the Wild Sweetgrass*, SpringerBriefs in Plant Science,
DOI 10.1007/978-1-4614-5903-3_5, © Robert J. Dufault 2013

Fig. 5.1 First planting of large, vigorous flowering sweetgrass at CREC on October 1991

is that these plants have been fertilized and watered probably heavily, rendering them luxurious but lacking strong fiber. Basketmakers have been very dissatisfied with this "commercial sweetgrass" for making their baskets. It is imperative that sweetgrass destined for baskets be grown very differently from ornamental sweetgrass that is grown for beauty only (Fig. 5.1). This will be discussed in the following chapters.

It should be noted that if sweetgrass is overfertilized, this situation is not permanent and can be reversed. In the above case, approximately 2 months later, the sweetgrass reverted back to a tighter "thread," and by the following year (after practicing benign neglect), the leaves were very strong and did not shatter upon pulling. Luckily, synthetic nitrogen fertilizers rapidly leach from soils after rainfalls and do not persist for more than a couple months. Fertilizing with organic composts should be avoided since their nutrient release is very slow, and if sweetgrass is given chicken, cow, or other rich composts, the sweetgrass may stay brittle for years.

"Ignorance is bliss." Whoever said that must have been ignorant. Despite failures and blind alleys, I was still very excited about working with sweetgrass after Keith retired, and I heard about the Palm Key failure (but at that time, I had no input to analyze why it failed until years later).Growing large acreage of sweetgrass was a challenge! Virtually "nothing" was known about its culture and I could "blaze new trails."

Low Country Sweetgrass Preservation Society

Another attempt at creating a local organization that would enhance the health and tradition of sweetgrass basketry was initiated by Joseph Riley, mayor of Charleston, SC. The goals of the society were as follows:

1. To help acquire land to grow sweetgrass
2. To work toward acquiring a comprehensive survey of sweetgrass habitats from McClellanville to Edisto Beach, SC
3. To gain permission to allow basketmakers to enter private lands to harvest sweetgrass
4. To work toward the cultivation of sweetgrass inland (Fig. 5.1)
5. To develop better public awareness of the sweetgrass tradition and challenges for survival and search for donors to the cause

The first meeting was December 3, 1991, and members included a group of people from the mayor's office, Mary Jackson, Robert Dufault, Keith Salvo, and others. We discussed the goals and work that needed to be done. But this group fragmented soon after and nothing was accomplished except some interested people made new contacts, and over the years, they offered assistance in various ways.

I continued to keep in contact with Mary Jackson and she became my mentor. Mary taught me so much about the plant, what the plant needed to "feel" like to be considered high quality, and the tradition of making baskets. It was agreed from the onset that I would seek opinion from the Mt. Pleasant Basketmakers' Association in any kind of endeavor or public relation activity. There is a history of exploitation of the entire Basketmaking Community for a variety of reasons, and I, for one, was not going to be in that group of people. If anyone wishes to "help" the situation, asking permission and/or keeping the community informed and asking their opinions was a paramount requirement.

> As for my motives, all I ever wanted was to end the shortage of sweetgrass and use my knowledge as a horticulturist to help without expecting anything in return except a feeling of achievement. And to this day, that motive has never changed.

Research is fueled by funding and so much is needed to be learned about growing sweetgrass. During these years of initial research, I had the opportunity to meet and get to know Margaret Davidson, executive director of the SC Sea Grant Consortium, NOAA. Margaret had become involved with the issues facing the basketmakers for a few years, and we had discussed the possibility of gaining funding through her agency. Upon her counsel, I wrote a proposal to cover a 4-year period. The main objective of the proposal was to develop a full cultural system for producing quality sweetgrass on inland soils. Specific subobjectives included the following: (1) to identify fertility regimes; (2) to characterize superior quality sweetgrass ecotypes from wild plants collected in SC, GA, and FL; (3) to develop low-input weed control techniques for plantations; (4) to determine spatial dynamics (plant populations and planting schemes); (5) to develop propagation techniques for field transplanting; and (6) to cooperate with Mt. Pleasant Basketmakers' Association on quality determinations. This proposal went forward for review but was rejected for funding although most of the reviewers rated it as "good to excellent." Competition for grant money was "very stiff," and unfortunately, my proposal did not make "the cut." So for the years that passed and submitting many proposals, the reviews of these proposals often would give me great accolades for attempting to problem solve with sweetgrass and "slaps on the back," but when it came time for funds to be granted, there was mostly silence. I did manage to receive small grants from gracious benefactors along the way, which will be mentioned, and again, my thanks extended for their generosity.

Chapter 6
Sweetgrass Utopia in the Southeast: Little St. Simons Island

As a horticulturist, the best way to learn about the cultural needs of a plant is to observe the plant growing in its natural habitat. Off the coast of St. Simons Island, Georgia, lies the small private island named Little St. Simons Island (LSSI) (Fig. 6.1). LSSI is secluded, serene, and filled with native flora and fauna and covers approximately 10,000 acres of rich tidal creeks, marshes, shrub land, maritime forest, and seven untouched miles of beaches.

And I will add LSSI is the home of the largest natural sweetgrass habitat I have ever seen, composed of many square miles of solid sweetgrass fields of the greatest beauty!

Luckily, Mary Jackson (Fig. 6.2) invited me to accompany her in June 1992 to visit LSSI with Mark Wexler, editor of National Wildlife Magazine, who was doing an article on sweetgrass basketry (Wexler 1993).

Little St. Simons Island History

During that short visit, I learned more about sweetgrass than I could have possibly learned in other manner. Virtually untouched by man, LSSI is the most remote and uninhabited of Georgia's barrier islands. The island's first owner was Samuel Ougspourger, a Swiss colonist from South Carolina, who purchased the island from King George II, in 1760, and eight years later sold it to his grandson Gabriel Manigault. Today, the island is owned by the relatives of Philip Berolzheimer, a wealthy New Yorker who acquired the island in 1908. Philip Berolzheimer (1867–1942) was born in Fürth, a town in Bavaria, Germany, where his grandfather Daniel was a pioneer pencil maker. Daniel had established a market for pencils in the USA as early as 1830. Philip immigrated to the USA in the late 1880s. Philip served as treasurer of the Eagle Pencil Company, and in 1908, the Eagle Pencil Company

Robert J. Dufault, *Stalking the Wild Sweetgrass*, SpringerBriefs in Plant Science,
DOI 10.1007/978-1-4614-5903-3_6, © Robert J. Dufault 2013

Fig. 6.1 Miles of sweetgrass grow profusely on Little St. Simons Island off the coast of Georgia. The greatest concentration is on the north end of the island on sand deposits of the Altamaha River as it exits to the ocean

purchased the 10,000 acre LSSI from the Butler family. The company's intention was to harvest the island's many red cedar trees for pencil production, but analysis soon determined that salt and wind had damaged the trees, making their wood unsuitable for pencil manufacture. A few years later, Philip visited the island, fell in love, and bought it from the company. The Berolzheimer family began to use LSSI

Fig. 6.2 Mary Jackson
holding a large bundle of
long, high-quality
sweetgrass harvested on
LSSI. Notice the large
sweetgrass field with wax
myrtle bushes in the
background

as a vacation destination, ultimately building a hunting lodge there in 1917. Throughout the 1920s and 1930s, the lodge was enjoyed by the family and their guests as a retreat for hunting and other recreational activities. In 1979, the Berolzheimer family converted their island residence into an expanded lodge to accommodate a maximum of 30 overnight paying guests.

An Amazing Sweetgrass Habitat

Miles of sweetgrass, amazing to behold but why was there so much was the logical question. The map above shows how LSSI hugs the coast of St. Simons Island. LSSI is bounded by the Hampton River on the south and the Altamaha River on the north. I learned from the LSSI naturalist that "accretion of sand along the north end beaches is incredibly swift with miles of beach added within a decade." The sweetgrass fields are especially common on the north end of the island. Subsequently, the Altamaha River rapidly deposits so much sand on that end and the first plant to colonize these new beaches is sweetgrass. The accretion is so swift, no dunes are formed, and sweetgrass starts colonizing this accreted land just beyond the high-tide mark on flat ground. As one drives away from the beaches, old primary dunes are apparent with plentiful sweetgrass, but then wax myrtles begin to colonize just beyond the interdune meadows which are characterized by flowering weeds, grasses, and woody plants (Fig. 6.3).

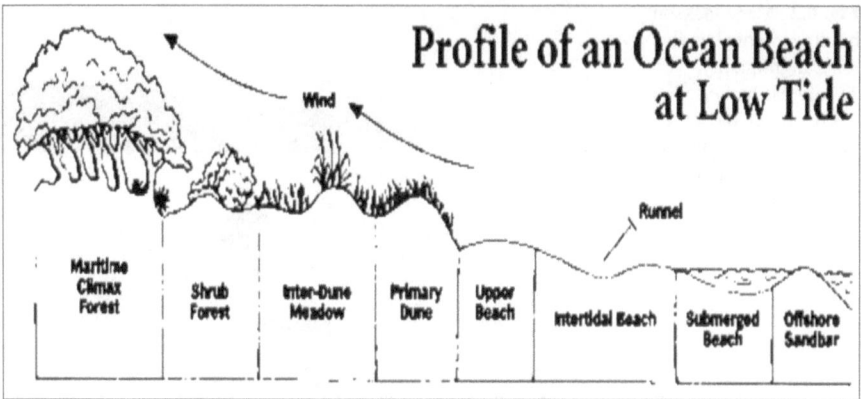

Fig. 6.3 Progression of plant types from the beachfront (*right*) extending into the forest on coastal barrier islands (*left*). Sweetgrass's natural habitat is usually behind the primary dune and sporadically up to the shrub forest

Eventually more aggressive woody plant species compete with the sweetgrass and "push" them out of that habitat (Fig. 6.4). Traveling further, the shrub forest gets thicker and the maritime climax forests "take over" and push out the shrub land, and the result is the creation of forest land with pine, wax myrtle, red cedar, and various oaks cloaked with vines. Animals, such as birds, snakes, rabbits, and deer, find shelter in this habitat with snakes especially liking the shade under sweetgrass plants. The habitats of LSSI sweetgrass are very different than any others I have observed in the Charleston area. The expansiveness of the sweetgrass fields was extraordinary, extending inland for possibly a quarter of a mile. Usually, sweetgrass grows in narrow bands in the interdunal areas just on the beachfront. On LSSI, sweetgrass was observed growing near the high-tide mark on the beach. Plants were also observed growing almost into tidal marshes with roots probably growing into brackish water (Fig. 6.5).

Sweetgrass was also observed growing on top of sand dunes potentially 10 feet above the normal ground surface (Fig. 6.6). Plants were excavated in many locations on LSSI and it was observed that sweetgrass is growing on top of dunes but, at one time, was growing closer to the ground, but with drifting sand, the plants can "move up" in the dune. I found evidence of lower buried primary root systems and another secondary root system much higher over the old root system now buried under drifting sand. Because of this growth characteristic, sweetgrass is an excellent sand-trapping plant that should be widely planted on renourished beaches to help stabilize dunes. Sweetgrass certainly is a xerophytic plant, can tolerate extreme drought, but equally tolerates brackish saltwater intrusion.

Fig. 6.4 Expansive natural groves of sweetgrass exist on Little St. Simons Island, growing in association with large wax myrtle trees (shrubs in background). Sweetgrass plants growing closest to wax myrtles are usually very long and highly desirable for basketry

On a second trip to LSSI in September 1992, I dug exceptional sweetgrass specimens to separate into smaller plants and propagate at the CREC. I also took soil samples under these plants for analysis to "learn" what soil nutrient values may correlate with excellent sweetgrass quality. The soil types at the beach were all characterized as pure sand. The results of the soil tests would provide guidance on inland cultivation, and these will be discussed in a later chapter comparing the soil tests from sweetgrass fields both cultivated and natural.

Visiting LSSI was the best education I could have received about learning what sweetgrass requires and what it tolerates. In cultivation, one decision that needs to be made is how far apart to space the plants. What better guidance to answer that query was to observe that the LSSI plants naturally spaced themselves out about 5–6 feet equidistant from their neighbors. Nobody planted these

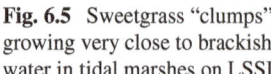

Fig. 6.5 Sweetgrass "clumps" growing very close to brackish water in tidal marshes on LSSI

plants yet they "seek" their own spacing, and potentially weaker sweetgrass plants growing nearby "lose" dominance as the plants continue to mature and grow wider as the years pass (Fig. 6.7). Sweetgrass plants do not send out underground runners and so all the sweetgrass seen on the beaches developed from seed. Although a single sweetgrass plant produces literally tens of thousands of seed, the success rate of any one seed producing a mature plant is infinitesimally small! In cultivation, however, "volunteer sweetgrass" plants are common since water and nutrients and human intervention increase the chances that volunteer seedlings will survive to maturity, but few do survive the "hoe." It is assumed that sweetgrass "in the wild" produces its load of seed in fall and the plants naturally "shatter" or disperse their seeds to the wind about end of November. Since sweetgrass is usually found in low spots on the beachfront, apparently seed blows into the interdunal areas, drifting sand covers the seed and winter rains may help spur slow germination in spring.

Fig. 6.6 Sweetgrass clumps growing near the top and the sides of sand dunes. Notice pine tree seedlings "moving in" and eventually will replace sweetgrass in this environment

Fig. 6.7 Sweetgrass clumps "seek" their own final distance from other plants in the "wild," spacing themselves about 5–6 feet apart naturally

Chapter 7
Getting to Know the Basketmakers

Mt. Pleasant Basketmakers' Association

By 1992, I had become invested in conducting research on sweetgrass. I was totally enthralled with the crop and had made a commitment to try my best to learn how to cultivate this crop and, in the future, reduce the supply issue. After 3 years of working with Keith Salvo and Mary Jackson, I learned enough about sweetgrass culture to share what I knew with the basketmakers. In conversations with Mary Jackson, I was invited to come to a Mt. Pleasant Basketmakers' Association meeting to talk with the basketmakers and introduce myself. It was also my goal to organize a Sweetgrass Culture Workshop to spend a day in lectures, greenhouse, and the field conducting some "hands-on" exercises to pass on what I knew about sweetgrass. On the evening of August 25, 1992, I attended the association's meeting and met the following basketmakers: Jeannette Lee, Joseph Mazyck, Annie Scott, Marguerite Middleton, Marie Rouse, Henry Wigfall, Harriett Brown, Florence Mazyck, Sue Middleton, Adeline Mazyck, Elizabeth Mazyck, Elizabeth Kinlaw, Linda Blake, Mary Choice, Leola Wright, Marilyn Dingle, and Albertha Robinson. I explained who I was, where I was raised, and my work experiences in horticulture and vegetable production. I mentioned that I felt I was an artist too, enjoy creating many handicrafts, and how I related to the art and craft of basketmaking. I also mentioned that my skills with plants could help unlock the mysteries of growing sweetgrass as a domesticated crop, but it will take more time to understand the best production system for the crop. This situation with sweetgrass is a common situation with new vegetable crops being introduced into any region. I stressed that there were great challenges with sweetgrass that needed to be solved starting from the simple act of finding enough plants in the wild to plant in inland locations.

My emphasis would be on planting in inland locations and not the beach and there were some aspects of growing sweetgrass that were particularly critical to solve such as the following:

Robert J. Dufault, *Stalking the Wild Sweetgrass*, SpringerBriefs in Plant Science, DOI 10.1007/978-1-4614-5903-3_7, © Robert J. Dufault 2013

Perceived Cultural Problems with Cultivated Sweetgrass

1. Weed control in inland locations will be a problem. There is a need to determine which herbicides can be used to chemically control weed infestations that are sure to occur.
2. What is the relationship among plant spacing, thread production, and strength? Can we plant very close and get maximum production in small fields and not reduce yield and quality?
3. The supply of sweetgrass already on the dunes is very limited. Removing these existing plants would further reduce the size of known natural sweetgrass habitats. Research is needed to learn how to start sweetgrass from separations of existing clumps into "mini transplants" and also starting plants from seed. In the establishment of new fields from plants taken from the dunes, it was unknown how small the clump separations could be to get good success. In other words, can the separations be the diameter of "dimes" or larger? In previous work, separations about the diameter of "fists" were very successfully transplanted, but this large size requires an enormous amount of clumps dug from the wild.
4. Should very small clump separations be grown for months in the greenhouse to "beef them up" before being field planted, and what size container is needed to maximize plant growth?
5. Is salt needed to grow sweetgrass? The dune area on the beach is very saline and we do not know if sweetgrass has a mandatory requirement for large amounts of salt or has sweetgrass evolved to tolerate high saline soil conditions.
6. What are the fertility requirements for sweetgrass that would increase fiber strength and thread length?

Although the funding for this work was very small, over the years, I practiced "serendipity research" and that is my term for trying something out in small unreplicated plots and hoping by "beneficial accident," I would learn something very useful. Through keen observation and through trial and error, much has been learned about sweetgrass that will be amplified in the Sweetgrass Biology section. After my meeting with the Mt. Pleasant Basketmakers' Association, a Sweetgrass Culture Workshop was planned and conducted at the CREC on Oct. 24, 1992. It was time to pass on what I had learned on to the basketmakers.

Chapter 8
Sweetgrass Culture Workshop: October 24, 1992

The workshop was a whole-day affair beginning at a.m. and ending about 4 p.m. Twelve interested individuals came that day including Adeline Mazyck, Ms. Snipe, Mary Alice Bostic, Florence Mazyck, Anna Bell Robinson, Harriett Brown, Albertha Robinson, Sue Middleton, Jeanette Lee, Leola Wright, Marguerite Middleton, and Cathy Townsend (horticulturist from Cypress Gardens, Moncks Corner, SC) (Fig. 8.1).

The topics covered included the following:

- History of sweetgrass research at CREC from 1989 to 1992
- Production of sweetgrass transplants (define quality, fertility, containers, media, age, and hardening)
- Field production of sweetgrass (soils, fertilizers, weeds, establishment, populations, and planting times)
- Tour of greenhouse containing 2,500 transplants destined for field planting at McLeod Plantation and a field plot of sweetgrass plots in bloom
- Transplant production lab (learn and practice clump splitting for bare root planting, mini clump splitting and planting techniques)

After the day was complete, I felt that I had made a contribution yet there was much to learn still about large-scale planting. My objective in the workshop was to teach the basketmakers to grow their own sweetgrass in their backyards or in cooperation with their friends. One aspect was clear to me after the workshop—I needed to plant a small basketmaker's sweetgrass garden that could be used as an example of what I would hope basketmakers could do on their own property (Figs. 8.2 and 8.3a–c). I had always felt that each basketmaker could take ownership of their own sweetgrass resource garden, and to this day, I feel that is a great "way to go" and what I would do if I was blessed with a sunny yard and the land. A 30 by 50 feet

Fig. 8.1 Workshop participants learning how to vegetatively propagate sweetgrass cuttings

Fig. 8.2 A small sweetgrass garden planted at CREC contained 7 rows about 30 feet in length with plants spaced within the rows about 1 foot apart. The ground was covered with landscape fabric to control weeds. This garden survived about 6 years before declining but could sustain the sweetgrass needs of 2–3 basketmakers

Fig. 8.3 Activities for the day included greenhouse propagation exercises (**a** *upper left*), a field trip on CREC farm (**b** *upper right*), and actual planting of sweetgrass in the field (**c** *below*)

sweetgrass garden would sustain the needs of probably 2–3 basketmakers. Eventually I did plant a sweetgrass garden and it existed for about 6 years before naturally declining. During those years, I allowed a few basketmakers to come harvest the sweetgrass to get their impressions of this idea. The idea was enthusiastically embraced.

Chapter 9
The Concept of Large-Scale Sweetgrass Plantations

After learning much from my sweetgrass cultivation at CREC, interactions with Keith Salvo and Mary Jackson, and the LSSI trip, I felt confident after 4 years (1989–1993) of trial and error, it was time to try large-scale planting. My thought was as follows.

What a great idea or fantasy—plant large tracts of sweetgrass with easy access by basketmakers on land donated by some benevolent entity, without any "strings," and grow free sweetgrass for all basketmakers to use. Basketmakers would not need to travel hundreds of miles and trudge miles along beaches looking for sweetgrass or ask permission to enter exclusive restricted beach communities to harvest anymore.

With much enthusiasm and not a lot of reservations, I attempted to plant large acreage of sweetgrass in two locations in South Carolina. In each instance, much was learned about the ease or problems inherent of making these plantations work. It is very important to review each attempt and evaluate the trials, errors, successes, failures, and lessons learned. And to this day, the idea of large-scale plantations still emerges occasionally especially with those who are "new" and do not know what has happened before.

Casually, farming sweetgrass seemed simple and straightforward enough. Using the example of establishing commercial vegetable production, in its most basic form, the farmer and the process of farming any new crop follows this pattern:

- Acquires the land (usually personally owns or rents)
- Plants transplants or seeds the crop
- Tends the crop (keeps out pests, weeds, and disease; fertilizes and irrigates)
- Harvests the crop
- Takes total responsibility for the crop's welfare
- Pays for the labor needed or uses his own "back" and then he "reaps what he sows"

Why not sweetgrass? In its natural habitat at the beach, nobody tends the sweetgrass and it grows all by itself. Why wouldn't sweetgrass do the same in some

Robert J. Dufault, *Stalking the Wild Sweetgrass*, SpringerBriefs in Plant Science,   47
DOI 10.1007/978-1-4614-5903-3_9, © Robert J. Dufault 2013

inland field? We assumed (erroneously) or wished that the latter would be the case. However, there is a vast difference between the scenarios of sweetgrass growing in its natural beach habitat and cultivating it inland and growing like a vegetable crop!

The obvious differences from vegetable farming processes above were as follows:

– Did not own land for sweetgrass.
– Did not have thousands of sweetgrass plants available to plant many inland acres.
– Did not have the equipment to tend the plants after transplanting in the field (tractors, sprayers, cultivation equipment, and irrigation).
– Did not have a defined labor force to "grow and tend" the plants.
– Did not know sweetgrass would be attacked by disease and insect pests in their inland locations.
– Could not regulate who harvested our cultivated sweetgrass and the fields had no security.
– Did not have a defined workforce who would take responsibility for maintenance of the planting.

Nevertheless, the odds were very much against success, but we attempted to cultivate sweetgrass, and we proceeded on a "wish and promise" and learned a lot in the process. For the sake of the future, it is so important to pass the lessons learned in these trials on to the next generation of "helpers" that are sure to come forward briefly in the future after they discover sweetgrass and want to offer some assistance.

Grand-Scale Experiments and Perceived Problems

It should be defined right at this point that the large-scale plantings at McLeod and Dill Plantations were grand experiments.

As all experiments go, when you initiate an experiment, you are not guaranteed a successful end result. In fact, you hypothesize what you think will happen, and many times, you learn eventually that your idea was proven false. When we started to cultivate sweetgrass, we certainly did not have any "tried and true horticultural wisdom." The questions we were asking were the following:

– Will the Basketmaking Community handle the care of these large plantings?
– Will domesticated sweetgrass even grow inland and will the quality be well suited for basketry on soils so different from their natural habitats?
– What pests will attack the plants? (We did not know that fire ants and black widows love to live in sweetgrass clumps!)
– How long will sweetgrass live in cultivation? Will it be decades like we assumed for sweetgrass growing on ocean dunes or a couple years?

– Can sweetgrass compete with weeds?
– Will the Basketmaking Community actually help with maintenance to tend the fields and regulate the harvesting of the sweetgrass?
– What will basketmakers think of the quality of this cultivated sweetgrass? (Too "hard" or not "soft" enough?)
– How many years will it take to get the plants up to harvestable stage of growth?
– Will snakes colonize the fields?
– Will basketmakers who did not contribute to the planting harvest these fields?
– How do you keep security of sweetgrass fields from those who would harvest and sell the grass?
– Will the Basketmaking Community take ownership and pride in this crop?

Chapter 10
Second Site of Large-Scale Sweetgrass Plantings: McLeod Plantation, James Island

The first real attempt of large-scale sweetgrass planting at Palm Key Resort by Keith Salvo failed due to improper soil fertility and lack of care. The second planting at McLeod was through my initiation. By serendipity, Mr. Lawrence Walker was a member of the South Carolina Agricultural Society, the second oldest agricultural organization that originated in colonial South Carolina. The "Ag. Society" was instrumental in the establishment of CREC in 1932 and has always maintained a very close relationship with CREC and its faculty. Mr. Walker also was the Executive Director of Historic Charleston Foundation. Established in 1947, the Historic Charleston Foundation is a nonprofit educational organization dedicated to historic preservation.

Historic Charleston Foundation owned McLeod Plantation at that time (Fig. 10.1). Mr. Walker was aware of the sweetgrass tradition, the shortage of sweetgrass, the need for a site to grow sweetgrass, and the need to preserve this ancient Low Country art and craft. Mr. Walker contacted me and offered the fields at McLeod as an experimental planting for sweetgrass, so with this offer came the opportunity we were looking for.

McLeod Plantation History

Located across Charleston Harbor just southwest of the city, McLeod Plantation encompasses 60 acres of fields and woods. McLeod Plantation is a vestige of the 17 plantations that once covered James Island and its fields were once toiled by enslaved Africans. McLeod is the last intact plantation on James Island with old slaves' quarters (Fig. 10.2) and is a historic landmark. Locating this field of sweetgrass in clear view of the row of old slave cabins was truly a testament of changed times.

Descendants of those who lived in those cabins are now using the same fields to plant sweetgrass that their ancestors toiled in to grow cotton. Their objective was to

Robert J. Dufault, *Stalking the Wild Sweetgrass*, SpringerBriefs in Plant Science,
DOI 10.1007/978-1-4614-5903-3_10, © Robert J. Dufault 2013

Fig. 10.1 McLeod Plantation house is the last intact plantation of seventeen plantations that once existed on James Island, SC

Fig. 10.2 A row of slave cabins on McLeod Plantation still exists where once the enslaved Africans toiled as in the same fields that sweetgrass was planted by descendants of enslaved Africans

preserve and enhance the future continuance of this African craft that sailed from Africa to the American colonies hundreds of years before.

In addition to its location and accessibility, the property features an antebellum plantation house. The area that is now known as McLeod Plantation first appeared on a 1695 map as a 617-acre plantation along the Wappoo and Stono rivers on James Island. In the early days, Sea Island cotton was cultivated there. In 1851, the plantation was sold to William Wallace McLeod, whose name it now bears. According to the 1860 census, 74 slaves lived in 26 cabins on the property. A number of these cabins and other outbuildings that supported the slave economy remain today as perhaps the single most striking feature of the property. The cabins measure about 20 feet by 12 feet and are of wood-frame construction on raised masonry pier foundations with exterior end chimneys. During the Civil War, McLeod Plantation served as Confederate Unit Headquarters, a commissary, and a field hospital until the island fell to the invading Federal army in the spring of 1865. When Confederate forces evacuated Charleston on February 17, 1865, Federal troops used the plantation as a field hospital and officers' quarters. The 54th and 55th Massachusetts Volunteer Regiments, composed of African-American soldiers, were among the units that camped at the site. The front parlor was used as a surgical theater and many Union and Confederate dead were buried at nearby Battery Means. There is also an old slave cemetery along the Wappoo banks in front of the house that was lost in time but "found" again in the 1990s when the land was being excavated for a fire station. It has been preserved now and left as a memorial. After the Civil War, McLeod Plantation became headquarters for the Freedmen's Bureau for the James Island district and in 1879 the McLeod family regained the property. In 1918 William Ellis McLeod began raising potatoes, asparagus, and dairy cattle. At his death in 1990, McLeod left the property to the Historic Charleston Foundation.

Objectives of the Grand Experiment: People and Science Based

With great hope, plans were laid to prepare a one acre block of land for sweetgrass planting. Unfortunately, there was not any farm equipment at McLeod or irrigation for any cultivated plants! So the immediate problem facing success was how to plant sweetgrass so it would be almost maintenance free. So, with enthusiasm, members of the Mt. Pleasant Basketmakers' Association and I toiled to prepare and plant 2,000 sweetgrass plants in an effort to eliminate the shortage. Large clumps of sweetgrass were dug from LSSI, separated into small "quarter" size, grown for a few months at CREC, and used in the first planting at McLeod. In June 1993, the first acre was planted (Fig. 10.3). The sponsors of these plantings included the SC Sea Grant Consortium, City of Charleston Department of Parks, Clemson University, and the Agricultural Society of SC. In May 1994, a second acre was added. The plant material for this second planting came from large plants dug at Kiawah Island, thanks to the generosity of the Kiawah Resort Associates and a grant from the

Fig. 10.3   On June 13, 1993, the first sweetgrass planting was made at McLeod Plantation with the help of 100 volunteers

Trident Community Foundation. The National Civilian Community Corps (NCCC) planted this second acre at McLeod and Dill Plantations, another remnant plantation on James Island, South Carolina.

Prior to planting each acre at McLeod, the entire field was covered with nursery landscape groundcover costing approximately $2,500 per acre. This material allowed the passage of water through the pores yet prevented weeds from coming through the fabric. The ground cover should last 10 years without deterioration. Ultimately, the weeds found their way through the planting holes and strongly competed with the sweetgrass. By fall 1993, the sweetgrass was flourishing (Fig. 10.4) and by fall 1994, the plants were lush and produced magnificent mauve pink flowers (Fig. 10.5).

This "grand" experiment at McLeod actually had two objectives:

1. People based: To determine if the Basketmaking Community could manage and take responsibility for the maintenance and harvesting this resource.
2. Science based: To learn how to grow cultivated sweetgrass and observe how it would perform in a large inland site which was very different from its natural habitat and assess production challenges.

Fig. 10.4 By fall 1993, the newly planted sweetgrass field at McLeod was flourishing

Fig. 10.5 The author, Bob Dufault, at McLeod Plantation examining sweetgrass in October 1995 when plants were in *full bloom*

Fig. 10.6 Eventually, weeds crept into the planting holes in the groundcover. Tenacious weed seed even germinated on *top* of the landscape fabric in decaying leaves. Crews of "weed pullers" could have rescued the fields to keep them clean

Results of People-Based Experiment

Initially, I felt we scored great success in this entire endeavor. The Basketmaking Community enthusiastically helped in 1993 with the production of the transplants and the field planting at McLeod. Also, it was proven that sweetgrass can be "tamed" and cultivated as a row crop; however, it must be "farmed" or cared for to be successfully grown.

For two years that followed, however, the maintenance of the fields was left predominantly to me and Clemson University CREC personnel. Although basketmaker maintenance groups were to be formed, only a few elderly men in the Basketmaking Community actually came to maintain the two acres of sweetgrass. By 1995, I transferred the responsibility of the maintenance of the field to the Mt. Pleasant Basketmakers' Association because maintenance was becoming a huge burden and weeds had managed to take hold of the field (Fig. 10.6) despite the ground cover.

By 1997, the task was too large for the Basketmaking Community to handle and the plantation was unfortunately abandoned. If this failure in maintenance did not happen, this resource could have thrived for at least 5–6 years. Ultimately after a few more years, Historic Charleston Foundation allowed the City of Charleston to move the plants to Lockwood Boulevard and they were planted along the Ashley River near

Fig. 10.7 Sweetgrass, originally planted at McLeod Plantation, was moved by the City of Charleston to a municipal planting along Lockwood Boulevard that borders the Ashley River. This picture was taken on April 2010, almost 17 years after these plants were planted at McLeod

the Coast Guard Center under a canopy of palmetto trees. Even in this location, weeds have competed with the sweetgrass, reducing their growth and quality (Fig. 10.7).

There were many horticultural lessons learned about how to grow sweetgrass inland, but after the demise of the McLeod planting, I was very skeptical that large-scale sweetgrass plantations will ever work until a dedicated and committed organization takes ownership and responsibility of nurturing the farming operation. In itself, maintaining large acreage of plants with a full-time staff is a very difficult task to accomplish since it requires great management skills, rules, regulations, time tables, equipment, schedules, and guardians to ward off any pests. We also had major problems by individuals who do not contribute to the work on the plantation yet felt entitled to harvest the grass covertly. The people-based experiment proved that large-scale plantations are too risky and the community cannot handle the maintenance.

Results of Science-Based Experiment

Horticulturally, the best way to learn about the risks of growing any new crop, whether it's a vegetable or a fiber plant like sweetgrass, is to grow large fields of the crop. In the small plots at CREC, the small nuisances and obstacles of production were masked, underrated, or easily fixed. But upscaled to large acreage, these nuisances quickly became overwhelming. The following points outline the major observations made and lessons learned.

Young Sweetgrass in Its First Year in Cultivation Has to Be Irrigated for Maximal Success. In its natural habitat on the beach, sweetgrass is a xerophytic plant (a plant that can grow in very dry conditions and is able to withstand periods of extreme drought). However, sweetgrass is only xerophytic when it is well established which begins in its second year after planting in cultivation. Transplanted sweetgrass should be irrigated during its first establishment year, and at McLeod in both years soon after planting, severe drought ensued as well as extremely hot summer temperatures. Water stress challenged the young plants and in the second year's planting, many died without timely rainfall and lack of any irrigation.

Sweetgrass Does Not Compete Well with Weeds. In a similar situation as just discussed, native sweetgrass growing in its natural habitat is not usually competing with massive weed infestations as it does in inland locations. The weed species in its natural habitat are totally different from inland weeds. Inland weeds species are extremely vigorous and present strong competition to cultivated sweetgrass (Fig. 10.8). Since the beachfront is such a harsh, dry, and intensely hot habitat, most inland weeds will not survive at the oceanfront. On the beach, occasionally an annual leguminous weed species will creep over sweetgrass, but these annual weeds die in the winter and the seed of the next generation has to sprout and compete with the large sweetgrass clumps. In the inland situation, each square foot of soil contains millions of dormant weed seed and that weed seed bank can survive for decades. In my 30 years of working with cultivated crops, there never has come a time when we ever depleted the weed seed bank in a field and did not have to use herbicides or cultivate biweekly to control weed infestations. My attempt to control weeds with landscape fabric in the sweetgrass plantings was a valiant attempt and clever idea, but the weeds were "smarter" than I was. Without manual labor to pull weeds out of the planting holes, the weeds ultimately won the battle. There were four problems with the landscape material.

Landscape Fabric Is Very Expensive for Large-Scale Plantations. In future plantations, chemical weed control (herbicides), tractor cultivation, and hand hoeing the fields would be mandatory.

Weeds quickly germinated immediately next to transplanted sweetgrass and began growing faster than the sweetgrass, producing extreme competition, and without labor to pull these weeds out, sweetgrass lost the battle.

As the sweetgrass clumps grew in diameter, the planting holes in the landscape fabric had to be manually cut much wider to allow the sweetgrass clump to grow larger in diameter over time. In some cases, the leaves would grow under the landscape fabric and the leaves would have to be pulled out from underneath. This manual cutting of fabric to make larger holes was accomplished in the second year at McLeod by the NCCC volunteers.

Poisonous Black Widow Spiders Infested the Microenvironment Under the Landscape Fabric Near the Plants. In the process of demonstrating to the NCCC volunteers how to cut the fabric to allow greater room for the sweetgrass to grow, during my demonstration, black widows emerged out of the expanded hole I just made!

Fig. 10.8 Weeds
germinated in the planting
holes and severely
competed with the
establishing newly
sweetgrass plant

Therefore, unlike snake problems in sweetgrass's natural habitat, black widow spiders were an inland problem with this landscape material.

Fire Ants Are Highly Attracted to Sweetgrass Clumps and Are a Huge Pest Problem. Fire ants are not found on the ocean front but are a ubiquitous inland pest. In the Southeast, fire ants are a major problem in yards and other areas that people live and congregate. After a year of growth, the sweetgrass clumps became large; the shade, moisture, and structure of the sweetgrass plants all provided a great attractant to the ants, and eventually, the fields were infested with these biting ants (Fig. 10.9). If a clump is harvested and the sweetgrass "pulled" as customary, it would not be unusual to have a few ants accompany the harvested leaves. Some people are so allergic that a single fire ant bite can send them to the emergency ward! So in domesticated cultivation, sweetgrass should be treated with insecticides to keep the ants out and prevent colonization. If fire ants are allowed to go untreated, they can ultimately kill the sweetgrass plants by throwing up huge mounds of soil into hills and eventually "chokes" the plant.

Fig. 10.9 A mature sweetgrass clump totally overcome by a fire ant mounds

Digging Existing Plants on the Ocean Dunes Is a Poor Way to Produce Transplants for Cultivated Sweetgrass Fields. All the sweetgrass for planting at McLeod was propagated by plants I removed from the natural habitats on LSSI and Kiawah Islands. These clumps had to be dug, transported back to the greenhouse, and separated by hand and the entire process took too much time. This process was laborious, reduced the natural, limited supply, produced a limited number of transplants, and required many clumps initially because large "fist size" separations survived better than small "dime size" separations. It was important in the future, if we intended to produce tens of thousands of plants for cultivation, we needed to learn how to germinate the seed and start new plants by seeding rather than clump separation. I had been trying and failing for years to germinate the seed and by serendipity, I discovered the secret to germinate sweetgrass seed (see in the "Biology of Sweetgrass" section).

Hill Culture Was Unnecessary for Cultivated Sweetgrass Production. In the Southeast, vegetable crops are placed on "hills" or ridged beds to increase water drainage away from the plants (Fig. 10.10a *top*). Ridges also warm up early in the spring by catching more sunlight as heat to warm the root systems of perennial crops. In the first year at McLeod, we used the hill culture system and found it was unnecessary and also it was extremely difficult to roll the landscape fabric over the ridges and anchor. In the second year, we used flatland culture (Fig. 10.10b *bottom*).

Years to First Harvest. Sweetgrass required at least two growing seasons in the field before the clumps were large enough to permit some harvesting of leaves. If done earlier than that, on these sandy soils, there could be a risk of dislodging the plants

Fig. 10.10 Sweetgrass was planted on top of "ridges" (**a** *top*) to increase water drainage at McLeod Plantation in the first planting. This technique was unnecessary and increased the toil of spreading landscape fabric. Sweetgrass planted in the second planting at McLeod was on flat ground (**b** *bottom*)

and pulling them out of the ground with premature harvesting of pulled sweetgrass.

Complaints of "Too Stiff" Sweetgrass and Too Rough on the Hands. Some basketmakers complained that the McLeod sweetgrass was too wiry and difficult to work with and tough on their hands, yet others felt it was just fine for their use. As I later learned and this is discussed in the "Biology of Sweetgrass" section, basketmakers have different preferences for texture of sweetgrass they use in their baskets. Botanists dispute the actual botanical name, yet to the basketmakers, there are "soft" and "hard" sweetgrass plant types. My objective at McLeod was the "hard" type because I had found excessively fertilized sweetgrass "broke apart" in the harvester's hands and it was totally unacceptable for basketry. The complaints voiced were related more to the situation of sweetgrass preference for the two types and a lack of communication of what my grower's goal was for quality.

Sweetgrass Is Susceptible to Diseases and Short Life Span in Inland Cultivation. A plant placed in a radically different habitat, sometimes becomes susceptible to diseases never encountered in its natural habitat. We noticed that an occasional plant would die in the field and this was disturbing. Samples of afflicted plants were assayed for pathogens and we found that sweetgrass was suffering from a "leaf spot" disease which could have been one of the following: *Curvularia* sp., *Bipolaris* sp., and *Drechslera* sp. Although there is nothing labeled by law for disease control on sweetgrass, in an emergency situation, the fungicide Manzate may control these pathogens. We also noticed in the years to follow, in all our plantings, the sweetgrass was a "short-lived" perennial and survived for 5–6 years then declined and died. This demise would require replanting fields to replenish those that perished over time. It is unknown in their natural habitat how long sweetgrass plants live, but it is suspected that they may live for decades.

Chapter 11
Third Site of Large-Scale Sweetgrass Plantings: Dill Sanctuary, James Island

Dill Sanctuary History

In conversations with Mary Jackson in the early 1990s, we attempted to calculate the total acreage that would be needed to sustain the sweetgrass needs of all basket-makers in the Low Country. After some deliberations, we estimated that upward of 40 acres would be necessary. So our small acreage at McLeod Plantation in the early 1990s would hardly make a dent into the grand needs of the entire community. So our quest was to find more land, preferably owned by a public agency or a humanitarian organization, that would allow use to grow sweetgrass for free and let the community organize itself to manage the distribution and harvesting of the crop. The McLeod planting was publicized in popular media such as CNN, SC ETV, NY Times, Philadelphia Inquirer, The State, Post and Courier, National Geographic Magazine, Island Magazine, and Carolina Camera television show. As a result, a variety of people approached me to offer land for sweetgrass planting, but their desire was to sell it to the basketmakers.

Dr. Will Post, researcher with the Charleston Museum, contacted me about using Dill Sanctuary on James Island, which is about five miles away from McLeod Plantation (Fig. 11.1). The Charleston Museum was founded in 1773 while South Carolina was yet a British colony. Now a modern 501(c)(3) nonprofit organization, the museum is accredited by the American Association of Museums. Dill Sanctuary was acquired by the museum in 1985. The Dill Sanctuary encompasses 580 acres and contains numerous cultural features including three earthen Confederate batteries and prehistoric, colonial, antebellum, and postbellum archeological sites. Dill Sanctuary has been protected for purposes of preservation, wildlife enhancement, research, and education and is used only for museum-sponsored programs. Dill Sanctuary provides assorted habitats for wildlife and nesting sites for both migratory and resident birds.

In fall 1993, I meet with John Brumgardt, Charleston Museum Director; Brian Varnado, Assistant Director; and Will Post about the idea of planting sweetgrass at

Robert J. Dufault, *Stalking the Wild Sweetgrass*, SpringerBriefs in Plant Science, DOI 10.1007/978-1-4614-5903-3_11, © Robert J. Dufault 2013

Fig. 11.1 Dill Sanctuary was the site of a plantation in the eighteenth century and has panoramic exposure to the intercoastal waterway and excellent climatic conditions for sweetgrass plant growth

Dill Sanctuary. My desire was to grow sweetgrass for the basketmakers' use, without any cost to them for these materials and this would be a humanitarian activity between Clemson University and the Charleston Museum. Dr. Post was also interested to determine if some bird species would use the sweetgrass for nesting. I was invited to use the sanctuary as a site of sweetgrass cultivation. I remember my meeting with Dr. Brumgardt. It was fairly rigorous and confrontational and my intentions/their commitments actively questioned. I remember leaving the museum office feeling like I had been through a PhD defense! This feeling should not have been dismissed since a couple years down the road, our mutual objectives for doing the sweetgrass work at Dill Sanctuary diverged drastically from what was discussed at this meeting, which led to an impossible impasse.

Results of Dill Plantation Field Experiment

The funding for the first field planting at Dill Sanctuary was provided by the Agricultural Society of South Carolina. The field selected was near the intercoastal waterway and an ideal environment for sweetgrass; however, it was learned much later that this field was an important archeological location and, at one time, was a dump site for plantation outbuildings. The field was littered with pottery chards and

Fig. 11.2 Sweetgrass planted at Dill Sanctuary by the end of summer 1995. Sweetgrass weed populations were controlled with herbicides and tractor cultivation, but the site did not have irrigation and we depended on natural rainfall for moisture

glass that may have dated back to the eighteenth century. Strangely, the museum later requested that no pottery chards could be picked up by visitors to the field since this location was considered tampering with artifacts. This aspect created much confusion and tension since the chards were very interesting to behold! This field should never been used for sweetgrass in the first place if it was archeologically sensitive!

In this first planting of approximately one acre (Fig. 11.2), we did not use landscape fabric to control weeds, but Charleston Museum retained a caretaker at Dill Sanctuary, who was to help us by cultivating the planting. In 1995, a large two acre block was planted at Dill Sanctuary (Fig. 11.3) and we had very high hopes that these plantings would be very successful. Unfortunately, the instructions I gave on many occasions were ignored and weeds flourished. Weed control required hand labor besides tractor cultivation since weeds grew between the plants and the cultivation could not sweep those away. Hence, without a labor force, CREC supplied the labor to hoe this field, and once again, we were faced with maintenance problems without much help from the Basketmaking Community. History was repeating itself similar to McLeod Plantation. The sweetgrass grew luxuriously over the next year but the weed control was becoming a large issue. With the lack of cooperation at following instructions at Dill, I approached the administration of the Charleston Museum with my concerns about our work at Dill. Soon after, I was informed that

Fig. 11.3 In spring 1995, the second planting of two acres was planted on bare ground at Dill Sanctuary in a deep, fertile field near the river. This field was an excellent choice for sweetgrass. The NCCC volunteers planted this field with seedlings grown from seed collected at McLeod Plantation from plants originally dug on Little St. Simons Island, Georgia

the Charleston Museum changed their view on the value of the plantings. I was approached with the idea they wished to charge basketmakers a fee for the sweetgrass! They also stated that no one was to enter the property, even me, without their permission. With this change in objectives and restrictions, I withdrew my support of continuing this endeavor at Dill.

What was learned from this attempt? Horticulturally, sweetgrass can be grown excellently in rows with tractor cultivation and with herbicides. The field used was very close to the intercoastal waterway and had probably a much higher humidity level that enhanced the production of gorgeous sweetgrass. The soils were very deep, light alluvial sand deposits with a high water table. From the manpower part of production, despite these positive observations, astute and timely cultivation of the plantings was necessary and critical to success. If these fields were tended by a diligent farm manager, they would have produced high quality sweetgrass, without a single doubt. In 2005, I visited the first field planted with Mary Jackson and its vigor was very poor. I learned that it was subjected to yearly burning by the caretaker, which is NOT correct and should never be used on sweetgrass because it kills the growing points of the plants. Lastly, large-scale plantations will never work unless the community contributes its time to help grow the sweetgrass.

Reevaluation of Large-Scale Sweetgrass Plantations

After my experiences at McLeod and Dill Sanctuary, I took stock and reevaluated the needs and solutions to inland large-scale sweetgrass production.

It was clear that large-scale plantations are riddled with problems including:

1. No one took ownership or responsibility except the initiator (me).
2. We were plagued by inadequate maintenance except by a very small group of people.
3. No harvest control was possible and the fields were harvested by poachers.
4. No security of plantings from "sweetgrass poachers" was possible.
5. We did not own the land nor had any control of the resource and land benefactors can override any decision concerning the destiny of the planted crops.

I came quickly to the conclusion after those large-scale plantings that small-scale home/community gardens would be one of the best options to the supply issues, providing a basketmaker had suitable land to plant a 30 feet by 30 feet garden. Individual basketmakers or groups of basketmakers can plant suitable backyards with sweetgrass. Our small sweetgrass garden at CREC thrived and provided more than enough sweetgrass for a few basketmakers. Basketmakers that plant individual sweetgrass gardens take ownership of this resource as they would a vegetable garden. As with any garden, the owner has maintenance responsibility and harvest control, and the garden is secure within the owner's property. The advantages for a sweetgrass garden are to supply needed materials at the basketmaker's back door. Valuable time and money would be saved and independence gained from the long arduous trips to find sweetgrass on the beach. Also, why pay someone else for a crop that grows relatively easy, without much care after the first year except for weeding and controlling fire ants? The only aspect that is critical with these gardens is they need to be planted in areas with full sun all day long. Sweetgrass that gets partial sun or afternoon shade will not grow vigorously, will slump over, and will ultimately fail. No exceptions can be made for the need for full bright sunlight and minimal shade. Besides sweetgrass home gardens, there are other alternative ways to access sweetgrass.

Chapter 12
Alternative Ways to Access Sweetgrass

Dependence on Private Beachfront Communities to Allow Access to Sweetgrass

Over the years, the Basketmaking Community has worked with various beachfront communities to get permission to enter their property and harvest native sweetgrass growing along the ocean. Communities such as Kiawah, LSSI, and Dewees Island have periodically hosted the basketmakers. The advantages were simply to access, once again, ancestral sweetgrass habitats. The disadvantages are that some of these locations are distant and expensive trips or just plain inconvenient. Once there, locating sweetgrass and harvesting within a limited period of time can be stressful and arduous. In some locations, basketmakers only have access to "commercial sweetgrass" growing in cultivation and not the natural sweetgrass growing on the beachfront. Harvesting strictly commercial sweetgrass has disappointed the basketmakers since quality can be inferior to sweetgrass growing in the natural habitat. In some cases, basketmakers have had to donate a basket to the community as part of the agreement to enter the property.

Municipal/Commercial Plantings

Before the early 1990s, sweetgrass was not used in the landscape in the Charleston area and, in fact, it was relatively an unknown plant. I will take the credit for bringing this plant from the dunes to the yards and garden beds in the Carolinas. In the early 1990s, I met Mr. J. Guy, who is the owner of a Carolina Nursery (Fig. 12.1a *left*), a huge nursery in Monck's Corner, SC. Carolina Nurseries specializes in the production of native species for use in the landscape including sweetgrass (Fig. 12.1b *Right*).

Robert J. Dufault, *Stalking the Wild Sweetgrass*, SpringerBriefs in Plant Science, DOI 10.1007/978-1-4614-5903-3_12, © Robert J. Dufault 2013

Fig. 12.1 Carolina Nursery in Monck's Corner, SC (**a** *left*), was a huge operation that specializes in native plants as well as sweetgrass (**b** *right*)

Sweetgrass was observed by people in the "wild" and over time, there was great curiosity and potential demand for its beautiful mauve fall-flowering plant in the landscape. I spent a few hours with Mr. Guy and also provided him with a few mother plants that he could experiment with. I also shared with him the special way that plants could be started from seed versus separating large mother plants into small plantlets. I remember making Mr. Guy sensitive to the sweetgrass basketmaking tradition and asked him not to call any plants he grows as "sweetgrass" but as" pink muhly grass," and I believe he has always done that. It is a fact that many people adopt the word "sweetgrass" in some type of business venture, yet they have nothing to do with sweetgrass tradition or even the plant but they try to latch onto the goodwill that the tradition of sweetgrass basketmaking connotes. In any event, after my visit with Mr. Guy, sweetgrass was added to his plant inventory, and over the years, sweetgrass started to appear in municipal locations, golf courses, and in the home garden. Sweetgrass is now a very commonly used landscape plant throughout Charleston and used extensively in parks and subdivisions, and you would think the supply issue would be solved; however, it is not. In many cases this grass is on state property, city property, or private land. Consent is still required for picking this grass in these areas. Another problem with this "commercial sweetgrass" is that it is subjected to agrochemicals and its value in basketmaking can be destroyed by over-fertilization, lack of clump maintenance, or just not ideal planting locations.

Commercial sweetgrass, however, is not all bad, and many attempts have been made to incorporate sweetgrass in the landscape with a dual purpose of beautification as well as raw material for basketmakers. The most recent attempt to achieve this goal is the new Waterfront Park underneath the new Ravenel Bridge which connects peninsular Charleston with Mt. Pleasant. In fact, the City of Mt. Pleasant has dedicated a building, the Sweetgrass Pavilion, to house educational posters and videos describing the sweetgrass basketmaking tradition (Fig. 12.2). Basketmakers also sell their baskets in the Pavilion and they have a lottery drawing to determine which days of the week an individual basketmaker will be able to set up a display of baskets for sale. Commercial sweetgrass can be a definite resource for the basketmakers, but every attempt should be made when planting these dual-purpose plantings to isolate the sweetgrass plants from other landscape plants. Usually, commercial

Fig. 12.2 The Sweetgrass Pavilion at Waterfront Park in Mount Pleasant, SC, is dedicated to highlight the history, art, and craft of sweetgrass basketry

plants are irrigated and fertilized routinely to maintain their vigor (Fig. 12.3). Sweetgrass in cultivation should not be irrigated (except in the planting year) and never fertilized since the leaves get brittle and break easily. Most cultivated sweetgrass is planted on fertilized, irrigated land, and its quality becomes inferior, disappointing basketmakers.

One of the oldest examples of municipal sweetgrass available to the basketmakers for harvest is planted along the Ashley River just west of the Coast Guard Center (Fig. 12.4). After the demise of the sweetgrass planting at McLeod Plantation, the City of Charleston asked the Historic Charleston Foundation if they could transplant the sweetgrass along Lockwood Blvd. In the late 1990s, the sweetgrass that originated from LSSI and Kiawah Island was moved to this location and has been resident along the sidewalk under the Palmetto trees for over 15 years. This planting, however, has not been maintained properly and it is full of dead leaves and weeds. Basketmakers do harvest sweetgrass from this planting which is about a quarter of a mile long.

It has been my observation that most of the commercial sweetgrass plantings have not been renovated yearly. Commercial sweetgrass does need yearly maintenance in January. Sweetgrass clumps should be cut back to about 6 in. in height; this removes the dead leaves and allows only new vibrant growth to extend out from the clumps. Even though sweetgrass is a warm-season grass, it will continue to grow throughout the winter months very slowly. It has been my experience that sheared sweetgrass will produce new leaves in cold winter months and never really goes "dormant" like many other grasses do.

Fig. 12.3 Sweetgrass planted at Waterfront Park in Mount Pleasant, SC, shares planting beds with other landscape ornamentals receiving timely irrigation and probably agrochemicals

Restoring Natural Habitats Along SC Beaches

The best approach to reverse the supply issue, but the most difficult to achieve, is to restore natural habitats along the South Carolina coastline. Over the last decades, I have heard that idea proposed more as a dream situation since nobody really knew how to achieve such a massive task.

Army Corps of Engineers Dune Vegetation Shore Protection Project

By serendipity, the Army Corps of Engineers, Charleston District has been renourishing South Carolina beaches over the years, and as part of the process, they replant these renourished beaches with native plants. The Dune Vegetation Shore Protection Project started in 2005, installed sand fences on the renourished beaches, and planted native plants behind these fences to help catch and stabilize new dunes from erosion (Fig. 12.5 *right*). Sand dunes are the first line of defense against the damaging waves

Fig. 12.4 Oldest commercial planting of sweetgrass along the Ashley River on Lockwood Blvd. in Charleston originally planted at McLeod Plantation. After 17 years, it still shows vigor and survives although it could appreciate a good weeding!

of coastal storms. Dunes are very susceptible to wind erosion and depend upon beach grass and fencing for their growth and survival. A healthy dune system provides a sand reserve and source for nourishing a beach as it erodes. The fencing breaks up the wind column, transporting the sand; the grasses hold sand in place until needed by Mother Nature to reduce the rate of shoreline erosion.

The plants help lower the velocity of the wind at the dune surface, causing windblown sand to be deposited around the grass and fence line and be retained rather than blown away, and dunes are created. Sand fences and plants are used in conjunction to catch windblown sand grains, and eventually, sand dunes are formed.

In 2005, Mr. Tommy Socha with the Army Corps of Engineers contacted me about the Corp's work on renourishing the beach at Folly Beach. Tommy explained his knowledge and interest in sweetgrass and the sweetgrass tradition.

It was agreed that planting sweetgrass near renourished beaches could:

1. Help stabilize the dune system.
2. Be available for harvest by sweetgrass basketmakers.
3. Restore sweetgrass back to its natural habitat, and possibly in years to come, sweetgrass will naturally repopulate its ancestral home.

Fig. 12.5 Sweetgrass seedlings were planted at Folly Beach behind the first existing dune (*left*) and other sand-trapping plants (*right*) were planted behind sand fences on renourished beach on March 2006. Note: sweetgrass plants were planted approximately 50 feet inland behind the dunes

So after some conversation, we decided to collaborate on including sweetgrass with other plantings of sea oats (*Uniola paniculata*), bitter panicum (*Panicum amarum*), and marsh-hay cordgrass (*Spartina patens*) on Folly Beach renourishment project.

Tommy and I met and talked about the implications of this work. The sweetgrass plants would not be planted behind the sand fence on renourished beach but as far back as possible on state land behind the existing first dune (Fig. 12.5 *right*). In most areas at Folly Beach, there is only one dune before the beach houses and high tide. We wanted to place the sweetgrass in areas that would not be eroded by the ocean only in the most extreme cases and to preserve the new sweetgrass habitats for as long as possible.

In July 2005, I met Tommy at Folly Beach and we surveyed all access points (beach crossovers) to the beach to determine the best potential new habitats. We picked 13 locations, and each of these locations was accessible by a pedestrian crosswalk over the dunes which will allow easy access to the basketmakers. Folly Beach has many crossovers, yet only these 13 crossovers showed potential as suitable habitats for sweetgrass.

Fig. 12.6 Sweetgrass seedlings were grown at CREC, in 4″ × 4″ pots for planting at Folly Beach, March 2006

The aspects of the habitats for planting fields were

1. Lack of a lot of weedy vegetation.
2. Field should be "bowl-shaped" to plant in the bottom and rainwater would collect in the basin.
3. Some shade, especially planting under crossover boardwalks and near any wax myrtles.

Sweetgrass grows especially tall in its natural habitat growing close to wax myrtle bushes. Apparently the shade of the bushes provides, for a few hours of the day, a reduced heat load on burning sands and the plants stretch more for light. Usually, in areas where sweetgrass grows in association with wax myrtles, the land is lower or in swales and tends to accumulate rainfall since wax myrtles do not thrive on top of high sand dunes.

Sweetgrass Planting Logistics at Folly Beach, SC

I was contracted to grow 3,600 sweetgrass plants that would eventually be planted in these habitats (Fig. 12.6). Sweetgrass seed for this Folly Beach planting was collected from plants I originally transplanted at CREC but collected from wild plants taken from Kiawah and LSSI. The seedlings were grown in 4″×4″×4″ pots and fertilized with Osmocote slow-release fertilizer. In hindsight, in future plantings, long tubelike planting cells about 5 in. long and 2.8 in. in top diameter (38 cell star plug trays—TO Plastics #720562C) would be preferred because of better greenhouse efficiency (can fit more plants in an area) and the long root system is better to reach water on the dunes than a shorter cell pack. The seed were planted in December 2005 and the seedlings grown in a glass greenhouse until March 2006.

These plants were planted on March 14, 2006, on the beach by Earth Balance Ecosystem Restoration Specialists, North Port, Florida, who were contracted by the Army Corps of Engineers. Planting sweetgrass *back* to ocean dune habitats had been attempted by Karl Ohlandt at Dewees Island, yet this large-scale operation was new to me.

We really did not know the answers to so many important aspects of this endeavor such as:

1. Will the plants survive?
2. Will the weeds and heat kill the newly planted seedlings?
3. How long it would take for the plants to grow to a harvest size?

In cultivation in inland locations, it takes about two full growing seasons for the plants to reach large enough size for leaves to be pulled without overstressing the plants. Plants grown on the dunes, however, are subjected to greater environmental stresses, and it was expected that they would not survive or recover as well as inland cultivated plants. To insure better survival, the plants were planted into deeper holes. Augers were used to expeditiously drill holes into the sands about 6 in. deep and the 4″ plants were set into the holes (Fig. 12.7c). Before setting the plants, about one cup of slurry of Terra-Sorb hydrogel (Fig. 12.7a) and water was poured into the planting holes (Fig. 12.7b). Terra-Sorb is a super-absorbent, potassium-based copolymer gel that significantly increases the water-holding capacity of soil. The manufacturer claims it absorbs up to 200 times its weight in water, slowly releases it to nearby plant roots, and lasts for 5 years! Terra-Sorb will repeatedly absorb and release water for several years. About two tablespoons of Osmocote slow-release fertilizer were also placed in the hole before setting the plants and filling the holes. After planting, the sand around each plant was firmed by foot stepping to increase contact of the soil with the root system.

Folly Beach Sweetgrass Planting Locations

The following are the general Folly Beach locations (Fig. 12.8) where a total of 3,600 sweetgrass plantings are located.

Planting site	Access number	Street location
1	8[a]	8th Street West off of West Ashley Avenue
2	7[a]	7th Street West off of West Ashley Avenue
3	6[a]	6th Street West off of West Ashley Avenue
4	—[a]	West between house 611 and 613 West Ashley Avenue
5	5[a]	5th Street West off of West Ashley Avenue
6	3[b]	3rd Street East off of East Arctic Avenue
7	6[b]	6th Street East off of East Arctic Avenue
8	7[b]	7th Street East off of East Arctic Avenue
9	8[b]	8th Street East off of East Arctic Avenue
10	9[b]	9th Street East off of East Arctic Avenue
11	10[b]	10th Street East off of East Arctic Avenue
12	12[b]	12th Street East off of East Arctic Avenue
13	Washout[b]	North of 13th Street off of East Arctic Avenue

[a]South of Holiday Inn Hotel
[b]North of Holiday Inn Hotel

Folly Beach Sweetgrass Development Over the Years

The plantings were observed over the years to determine their survival success, vigor, and harvest status (Fig. 12.9). As stated before, the plantings were slow to "catch hold" and grow vigorously. In fall 2006, the plantings were all visited and evaluated for their stands and vigor. Unlike inland plantings, the plants had at least a 95% survival rate which was very pleasing, yet the size of the plants was still very small. Life on the dunes is harsh and the growth rate was deemed "very slow." Another year was allowed to pass, and the plants were evaluated again in 2007, and their growth was still very slow yet the plants still survived well. In August 2008, I invited Mary Jackson to view the plantings and get her opinion on the status of the plants. Together with her husband Stoney, we looked at all the sites and about half of these had matured to harvestable size and half very still undersized and considered "scrap." Scrap sweetgrass is inferior sweetgrass, not highly desirable, but in cases of lack of supply, they will be harvested

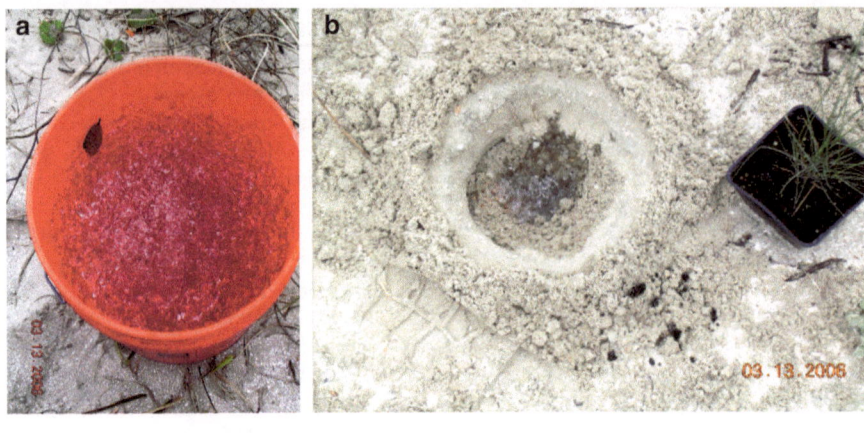

Fig. 12.7 Terra-Sorb slurry used to provide water supply in planting hole for many months after planting (**a** *top left*). Planting hole with Terra-Sorb and Osmocote in hole bottom (**b** *top right*) and worker drilling planting holes with an auger (**c** *below*)

for use. Mary suggested that the plants needed one more year to gain some maturity before they should be harvested. We also discovered that the quality of the plants was not the same at all locations. Some of the locations were considered very poor and may never mature to a quality status. The best locations proved to be those that were planted into the bottom of a swale that collected water. Again, we had a grand experiment and learned new things.

Another year passed and in May 2009, I invited Nakia Wigfall, Executive Director of the Sweetgrass Cultural Arts Festival Association, to view the plantings at Folly Beach and get her opinion. We visited every location and I provided Nakia with a map for her records and pass on the location of these

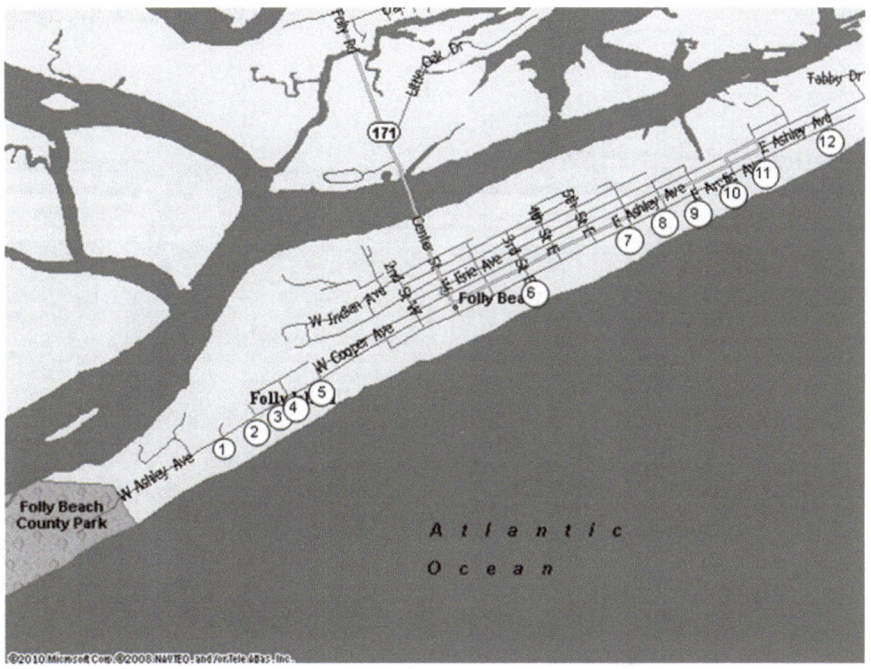

Fig. 12.8 Geographical locations of sweetgrass fields at Folly Beach, SC

Fig. 12.9 A typical sweetgrass field was planted behind the second dune at Folly Beach

Fig. 12.10 Bob Dufault (Clemson University) (**a** *top left*) and Tommy Socha (Army Corps of Engineers) (**b** *top right*) standing in a maturing sweetgrass field (**c** *bottom*) at Folly Beach in August, 2008

sites so basketmakers could eventually come and harvest sweetgrass. In the year between my visit with Mary Jackson and now with Nakia, the plantings had finally come "full circle" (Fig. 12.10a–c). Those locations that were poor

Fig. 12.11 Thirty-eight cell star plug tray (5″ long×2.8″ top diameter×1.7″ bottom diameter) used to grow Myrtle Beach plants (**a** *left*) and close-up of tops and root systems (**b** *Right*) of these superior plants for planting in ocean sand soils

had become poorer probably due to very poor sand (soil) conditions, yet some deemed inferior the year before finally grew "out" of any earlier problem and were good quality. Nakia was excited about what she saw and enthusiastic about seeing sweetgrass finally returned to its original home. The future looked brighter with these plantings, but it did take a full 3 years from planting for the plants to mature to a decent size, and in the fourth year (2010), they could be harvested by basketmakers.

Grand Strand Sweetgrass Project

Beach renourishment continued in fall 2008 in South Carolina and the Army Corps finished a much larger project, renourishing over 25 miles of beach from Garden City, Surfside, to North Myrtle Beach. Since the Folly Beach sweetgrass planting was so successful, 10,000 sweetgrass plants were planted over the 25 mile length of the renourished beach in 24 locations. The creation of these new habitats holds so much promise for the Basketmaking Community. With this planting, sweetgrass plugs were planted on the dunes using the new improved cell trays in March 2009 (Fig. 12.11).

Grand Strand Sweetgrass Planting Locations

Since it takes at least 3 years for sweetgrass to grow large enough to be harvested, it is expected that by 2013, these locations will be producing quality sweetgrass for Low Country basketmakers.

Planting site	Address	Additional information
Myrtle Beach reach one		
1	57th Avenue North	Laguna Keyes Condo
2	42nd Avenue North	
3	18th Avenue North	
4	12th Avenue North	
5	6th Avenue North	
6	2nd Avenue North	
7	21st Avenue South	
8	3513 South Ocean Boulevard	Spinnaker Condo
9	45th Avenue South	
Myrtle Beach reach three Surfside		
10	North End of Reach 3	Last House
11	1111 North Ocean	Brick Condos
12	10th Avenue North	
13	811 North Ocean	Sea Shadow
14	813 North Ocean	Ocean Club
15	4th Avenue North	Back Slope
16	4th Avenue North	Front Dune Swale
17	2nd Avenue North	
18	4th Avenue South	
Garden City		
19	House no. 893	Yellow House
20	House no. 927	Just short of 2nd walkway
21	House no. 973	
22	House no. 1005	
23	House no. 1323	Between public access 22 and 24

The Happy Ending to a Long Struggle

Of all the ways to replenish sweetgrass, I feel the ideal situation is to reestablish natural habitats to their original locations.

I will always be indebted to the Army Corps of Engineers and Tommy Socha for having the desire, ways and means, and foresightedness to accomplish this monumental task and help solve the supply issue in such a magnanimous way.

I hope in the future decade and those that follow that these plants returned to Folly Beach and Myrtle Beach area will naturally colonize more of the South Carolina coast and establish a wider distribution of sweetgrass colonies in areas adjacent to planted areas and once again allow the plants to "do their own thing" and also provide the Low Country basketmakers a copious supply of sweetgrass that is free for the taking without getting permission from anyone in the future.

Of all things that have happened since the supply issue became rampant, this, to me, is my "happy ending" I so desired for the Basketmaking Community.

Chapter 13
Sweetgrass Biology

Sweetgrass Botany: What's in a Name?

Sweetgrass is in the grass family technically called the Poaceae family. Debate has been ongoing for decades by botanical taxonomists about the correct scientific nomenclature to name the species of sweetgrass used in African-coiled basketry. When I started working with this plant in the late 1980s, it was accepted by most botanists that the genus and species was *Muhlenbergia filipes* M. A. Curtis. Alternative names are *Muhlenbergia capillaries* (Lam.) Trin. var. *filipes* (M. A. Curtis) Chapm. ex Beal and *Muhlenbergia sericea* (Michx.) P. M. Peterson. The genus *Muhlenbergia* was named after Dr. H. Muhlenberg, who wrote a work on American grasses in 1817. The genus *Muhlenbergia* has 152 species with 69 of these species native to North America (Gustafson and Peterson 2007).

Colloquial names include sweetgrass, dune hair grass, gulf muhly, pink muhly, and gulfhairawn muhly. Joseph Pinson and Wade Batson (1971) reviewed the status of *Muhlenbergia filipes* and made some very clear distinctions about this particular plant and how it deserved its own name because of its unique characteristics and very unique habitats compared to other closely related grasses. In the Low Country of South Carolina, three species of *Muhlenbergia* grow, and their identity can be confusing to the lay person yet examination of its botanical characteristics and their habitats provide evidence of distinctions in their relationship with the sweetgrass tradition. *Muhlenbergia capillaries*, *M. expansa,* and *M. filipes* are these species, yet their habitats differ as well as their floral anatomy (lemmas and glumes). The plant most typically used in sweetgrass basket construction and referred to by basketmakers as "hard sweetgrass" is *M. filipes*. Its habitat is in marginal habitats along coastal barrier islands usually in the swale between the interdunal beach areas. This "hard" sweetgrass is the longest and coarsest of the three *Muhlenbergias*.

Gustafson and Peterson (2007) have reexamined these three species and use the scientific nomenclature of *Muhlenbergia sericea* to refer to the previously named *Muhlenbergia filipes*. The name *M. sericea* is a much older name given to the species

Robert J. Dufault, *Stalking the Wild Sweetgrass*, SpringerBriefs in Plant Science, DOI 10.1007/978-1-4614-5903-3_13, © Robert J. Dufault 2013

by Andre Michaux who was a leading French botanist in colonial South Carolina in the 1700s. He was appointed by Louis XVI as royal botanist and sent to the USA in 1785 to investigate plants that could be of value in France. He traveled with his son Francois André (1770–1855) through Canada, Nova Scotia, and the USA. In 1786 he established and maintained for a decade a base in Charleston, South Carolina, from which he made many expeditions to various parts of North America. He described and named many North American species during this time and collected many plants and seeds to send back to France. At the same time, he introduced many species to America from various parts of the world, including sasanqua, tea olive, crepe myrtle, and maidenhair tree or ginkgo. Peterson (2003) rejected the newer name of *M. filipes* and opted to call it again by the older name given by Michaux. Therefore, *Muhlenbergia sericea (Michx.)* P. M. Peterson is now the accepted name of the basketmakers' sweetgrass type. Even though the scientific name has changed, the grass is the same "hard sweetgrass" that grows by the ocean's edge.

Basketmakers also use *Muhlenbergia capillaris* or as they refer to it as "soft sweetgrass." This soft sweetgrass does not grow on the ocean's edge but is a woodland habitat usually miles from the beach on the edge of pine forests. *Muhlenbergia expansa* grows in wet pine savannas and pitcher plant flatwoods far inland from the coast. Only *M. sericea* naturally occurs by the ocean's edge. The "soft" woodland sweetgrass (*M. capillaris*) is very fine in texture and usually much shorter than the "hard" *M. sericea* sweetgrass. Both species, however, have the pretty mauve flowering plumes appearing in September to October in Coastal South Carolina. Some basketmakers prefer the "soft" sweetgrass as it is easier on the hands and it also is easier to use in small objects, especially miniature baskets, hair ornaments, and jewelry.

Confusion is rampant in the commercial nursery industry that sells *Muhlenbergia* plant species. The commercial nursery industry has adopted the name *Muhlenbergia capillaris* apparently for both *M. sericea* and *M. capillaris* to the dismay of the scientific community. In fact, the preferred hard sweetgrass goes by many different variations of *Muhlenbergia* names; some spelled wrong in fact in the nursery trade! If anyone tried to buy hard sweetgrass from a nursery, invariably they will see the wrong genus and species name which may confuse them to what exactly they are getting.

Even though the general plant appearance of both *M. capillaris* and *M. sericea* are similar, the species can be separated by the botanical features, especially the flowering (awn) structure. An awn is either a hair- or bristle-like appendage on a larger structure or a needle-like element. Awns are characteristic of many grasses, where they extend from the lemmas of the florets. They often make up the hairy appearance of the grass's inflorescence. If you can imagine a head of wheat, all the bristles are awns (Fig. 13.1).

The awn on the upper glume of *M. capillaris* is shorter or absent; on *M. sericea*, the awn is very much longer (Fig. 13.2). Even more definitive to identification that must be stressed, *M. capillaris* does not grow on the ocean dunes, but it is woodland *Muhlenbergia* that grows on the edge of forests many miles inland.

Fig. 13.1 A definitive botanical feature to determine the correct species of Muhlenbergia is examination of awns. The hairlike projections on the head of wheat above are awns

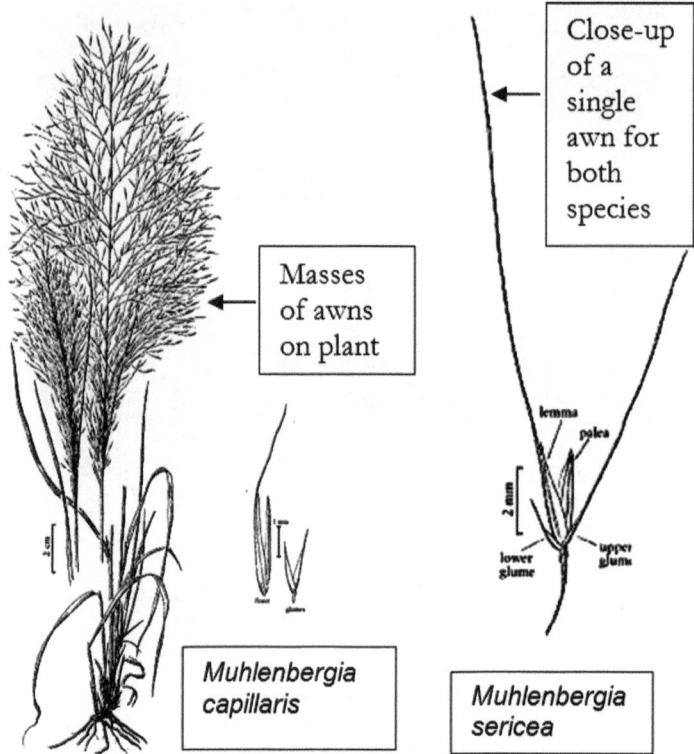

Fig. 13.2 Both *M. capillaris* and *M. sericea* are similar looking with mauve flower spikes, yet on the tiny pink future seed (*left*), the awn is much longer for *M. sericea* (*right*) than *M. capillaris* (*middle*) which may be very short or awnless in some cases (Source: Webmanual at Utah State, http://www.herbarium.usu.edu)

Chapter 14
Sweetgrass Horticulture: Environmental Considerations

Distribution in the USA and Habitats

Sweetgrass can be found in very narrow bands of a few 100 feet along the coast from the Outer Banks of North Carolina, through the Southeast States into the Gulf of Mexico States all the way to the southern tip of Texas (Fig. 14.1). Distribution information was extracted from herbaria, and it is expected that sweetgrass continues south through the coast of Mexico.

Sweetgrass is more widespread in Florida than in South Carolina. Throughout Florida coast, sweetgrass dominates the swales behind the front dunes (Meyers and Ewell 1990). The sweetgrass natural habitat has been described by Kurz and Wagner (1957) as the freshwater end of the salt marsh within 100 m of mean high tide. Sweetgrass grows in scattered clumps on relatively flat, but undulating areas bounded by dunes on the seaward side and by dense thickets of wax myrtles (*Myrica cerifera*) on the landward side (Pinson and Batson 1971). The natural process of plant succession for sweetgrass grasslands eventually would be to proceed during secondary succession to a maritime shrub and finally a maritime forest (Ohlandt 1992). Shrubs eventually will invade and replace wet maritime grassland; this was very obvious to me on my trip to LSSI in the early days.

Plant Longevity

In its natural habitat on the ocean dunes, it is unknown how many years sweetgrass lives. There is no published information about this enigma. In cultivation, we know that within about 6 years, the plants declined so much in vigor; they should be removed and replaced. We can assume though that in the future with our plantings at Folly Beach and on the Grand Strand, we will be able to determine how long the plants will live. It is expected that they will survive much longer along the ocean than inland since that is the environment they have evolved.

Robert J. Dufault, *Stalking the Wild Sweetgrass*, SpringerBriefs in Plant Science, DOI 10.1007/978-1-4614-5903-3_14, © Robert J. Dufault 2013

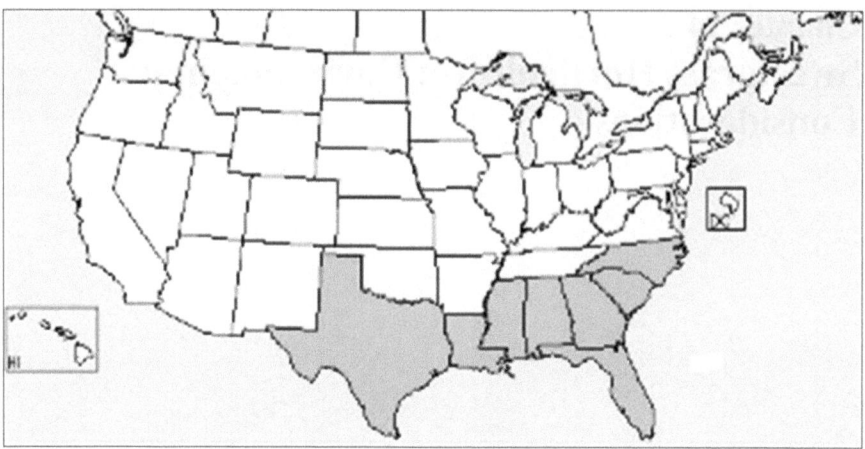

Fig. 14.1 Distribution of *Muhlenbergia sericea* in the United States. Note that within each state, the plants will only be found on narrow bands of about 100 feet wide along the coast only and not throughout the entire state as this map implies (Source: http://www.plants.usda.gov/java/profile?symbol=MUFI3)

Soils

If any factor can more affect the growth of a plant, it is the soil it grows in. On the dunes, sweetgrass cannot physically move to a better "neighborhood" and must scavenge for the necessary nutrients it needs in the place it grows. Sweetgrass has evolved on the salty interdunal swales on the Atlantic Ocean and the nutrient levels in those sandy soils suit the growth of sweetgrass. One of the greatest unknown questions about growing sweetgrass as a cultivated plant is "what soil conditions does the plant require to produce the ideal quality for basketmaking?"

In all our cultivation experiences, the greatest and primary problem facing us was finding ANY suitable land that would be free to use and easily accessible by the basketmakers. We did not have the luxury of accessing the perfect soil conditions that replicate the dune soils nor did we to replicate dune soils in the inland locations. We *assumed* that inland soils were well suited for vegetable or grain crops, would grow "good" sweetgrass. But that is not necessarily true. In fact many basketmakers have been vocal about trying to find land near the ocean for cultivation since they believe that quality sweetgrass needs to be planted on similar soils. They do have a very valid point.

In an attempt to understand the soil conditions that native sweetgrass grows on, during my trip to LSSI with Mark Jackson as mentioned above, we identified luxurious high-quality sweetgrass plants in various growing conditions and dug those plants for use in my inland locations. At the same time the plants were dug, I also took soil samples next to the root balls of those superior plants to analyze and then determine the soil fertility. The table that follows outlines the outcome of those samples. I have also included the soil test from McLeod Plantation which was considered to be well suited for vegetable crop production as a comparison to the beach environment.

The first aspect to discuss is the differences among sites at LSSI. Probably the most hostile location that any cultivated crop could endure would be on new beach accretion that seedling sweetgrass germinated and was flourishing on only within yards of high tide (see LSSI #1 in Table 14.1). Specifically, new beach forms very quickly as sand is deposited on the north end of LSSI from the Altamaha River, and sweetgrass is the first plant to rapidly colonize that area. Normally sweetgrass is found behind the first or second dunes, but in this LSSI location, beach accretion is so rapid, no dunes are present and sweetgrass germinates and thrives on flat beach sands just beyond the high-tide mark. Many soil elements were in the excessive range in location 1 including phosphorus, calcium, iron, and especially sodium. The soil pH and magnesium levels were in the acceptable range for most crops though, and other elements are considered low including potassium, manganese, zinc, and copper. The cation exchange capacity (CEC) was very high but this is not considered detrimental since this is a measure of the nutrient-holding capacity of the soil. A CEC of that magnitude is usually found in soils with high organic matter or high clay content, but this location is devoid of organic matter and clays. The soluble salts were very low and they are a measure of the soil's ability to conduct electricity; the higher the reading, the more salts the soil can hold. It is an enigma that these new beach soils have very high levels of calcium, sodium, iron, etc. and still have very low soluble salt measurements. It is my opinion that most cultivated plants would not be able to tolerate this level of fertility yet sweetgrass thrived on them.

In location 2, I sampled a field of sweetgrass that was growing on a sand plain about 1,000 feet from the beach. These fields could be described as old, mature, established, high-quality sweetgrass plant colonies. This sweetgrass was considered highly valuable and prized by basketmakers. The soil test results were similar to location 1 in that phosphorus, calcium, sodium, and iron were still very high with an acid pH, yet sweetgrass thrived.

In the last location 3, I sampled an excellent stand of sweetgrass growing within a few feet of tidal brackish water along an inland marsh. The organic matter level and soluble salt levels were still very low, and phosphorus, magnesium, calcium, and iron were high but sodium was considered at *extremely high* levels. Although inland, sodium must have accumulated in location 3 to a far greater extent than at the ocean's edge in location 1. Salt at this level would be highly toxic and herbicidal to most cultivated plants I am familiar with, and it is extremely apparent that sweetgrass was an extremely salt-tolerant plant. All the values for the three locations (1–3) were averaged to give a general idea of the soil status of high-quality sweetgrass growing in the wild.

In comparison to the soil test from McLeod Plantation, the organic matter, soluble salts, pH, magnesium, manganese, and zinc were similar to LSSI average. The CEC, sodium, phosphorus, and calcium were much lower at McLeod, while the potassium and copper levels were higher at McLeod.

The obvious question that needs to be answered is "does cultivated sweetgrass have to be fertilized to the level that exists in its natural habitat like LSSI to get

Table 14.1 Soil test results from natural sweetgrass habitats at Little St. Simons Island, Georgia, and cultivated sweetgrass site at McLeod Plantation, James Island South Carolina

Location	Beach or inland	Organic matter[a] %	Soluble salts[b] (ppm)	Cation exchange capacity[c]	Soil pH[d]	Phosphorus Lbs/acre	Potassium Lbs/acre	Magnesium Lbs/acre	Calcium Lbs/acre	Sodium Lbs/acre	Manganese Lbs/acre	Iron Lbs/acre	Zinc Lbs/acre	Copper Lbs/acre
Little St. Simons #1	Beach[e]	0.33 VL[f]	32 VL	14.4	6.3	2,164 V	20 L	93 M	5,290 VH	191 VH	13 L	217 VH	2.0 L	0.3 L
Little St. Simons #2	Beach[g]	1.04 VL	38 VL	9.5	5.8	1,076 VH	46 L	75 M	3,154 VH	108 VH	16 L	183 VH	6.1 A	0.2 L
Little St. Simons #3	Beach[h]	1.44 VL	166 VL	10.0	6.3	904 VH	214 A	373 VH	2,469 VH	464 VH	7 L	176 VH	5.4 A	0.4 L
Average Little St. Simons	Beach	0.94 L	79 VL	11.3	6.1	1,381 VH	93 L	174 H	3,638 VH	255 VH	12 L	194 H	4.5 A	0.3 L
McLeod Plantation, SC	Inland[i]	1.04 VL	38 VL	5.3	5.8	410 VH	323 H	76 M	858 A	66 H	14 L	46 VH	5.0 A	1.0 M

[a] Organic matter is accumulated decomposed plant and animal residues in soil, and the higher the number, the more productive the soil

[b] Soluble salts are expressed as the electrical conductivity of the solution. Salts can accumulate in soil and it may be difficult for plants to extract water

[c] Abbreviated CEC is a relative measure of the nutrient-holding capacity of the soil. Sandy soils have lower CEC, while organic soils have higher CEC

[d] Under normal conditions, the most desirable pH range is from 6.0 to 7.0 for most plants and 5.8 to 6.5 for grasses. A pH below 7 is considered acidic, 7 is considered neutral, and higher than 7 is considered alkaline. pH values outside of these ranges may cause certain elements to become more available or less and could incur deficiencies and/or toxicities

[e] Soil test of newly germinated seedlings on recently accreted beach. This test result indicated the optimal condition that favors sweetgrass seed germination in natural habitat

[f] VL low and very deficient for plant growth; L low and inadequate for plant growth; M medium availability; A ideal concentration for plant growth; H optimal, indicating ideal supply of nutrient for most crops; VH excessive, indicating more than necessary for optimal growth. A V indicates "very" and an extreme situation with a certain element that could seriously affect growth

[g] Soil test of old fields of quality sweetgrass growing about 1,000 feet from the beachfront. This test result indicated a maintenance level of soil fertility factors in a natural habitat

[h] Soil test of an excellent stand of sweetgrass growing near brackish tidal marsh. This test result indicates an ideal soil fertility status for excellent-quality, long sweetgrass growing near a high water table

[i] Soil test of field used to plant sweetgrass at McLeod Plantation on James Island, SC

high-quality sweetgrass?" The answer is I do not know. Especially in the case of sodium at LSSI (which would be toxic for most crops and sodium never is high in cultivated soils in Carolina), does sweetgrass have an obligate need for sodium to grow to a state of high quality? The answer is still I do not know. I feel it is very good to understand now the soil status in sweetgrass's natural habitat yet research is needed to determine the effects of sodium on sweetgrass quality and growth. In cultivation, sweetgrass grows "well" without additional sodium, but nobody I have ever known has used salt on sweetgrass in cultivation. It has been my experience though to put up to 2,000 pounds of salt/acre on asparagus fields, another salt-tolerant crop. In the case of asparagus, the yields of spears were boosted and the incidence of *Fusarium* disease was suppressed with salt. There is no reason to doubt that sweetgrass growth and its longevity improve with salt additions in cultivation. It can be assumed that inland diseases and weed would be suppressed with the addition of salt to cultivated sweetgrass fields, but this needs to be researched.

Heat/Drought Tolerance

Based on the hostility of the soil conditions, the pure sand nature of the ocean dunes, and the intensity and level of heat these soils absorb, sweetgrass still thrives on the beach and it can be considered a very heat- and drought-tolerant plant. Sweetgrass plants are blessed with huge, densely populated root systems that go very deep into the soil. This anatomical characteristic is one of the major reasons why sweetgrass is so drought tolerant, a prolific root system. The great number of roots gives the plant the ability to "gather" water and nutrients from a vast area around the plants and store the water for times of need. Sweetgrass also has very thick fleshy roots very similar to asparagus crowns, and these roots store sugars for plant metabolism needed in times of stress. The function of the leaves is to "catch" sunlight and carbon dioxide and convert them in the process of photosynthesis into oxygen and sugars. The leaves are the "sugar factories" and the roots are the "warehouses" where the supplies are stored until the plants need them for growth and development. In cultivation though, in the first year after planting, sweetgrass does need and depend on irrigation to survive and "take hold." It is not until the second year that sweetgrass will depend solely on natural rainfall to supply its water needs.

Pest Pressures

In its natural habitat, sweetgrass is one of the first plants to colonize the dune swale locations. There are a few beach weed species that can grow rampantly and choke out sweetgrass. In the succession of plants to colonize a beach and as the beach accretes, eventually wax myrtles and other woody shrub-like plants overtake sweetgrass locations, compete with the sweetgrass, and become the dominant species. In relation to diseases, depending on the location of the sweetgrass stands, I have noticed that the leaves of sweetgrass can be severely damaged and burnt by salt spray or by the potentially high salt content in the soil. Although their quality for basketry may be lost, these plants still tend to survive, having evolved in this location from time immemorial. In cultivated fields, sweetgrass plants eventually decline over a 5 year period. We have not done a complete sampling of the diseases that affect cultivated sweetgrass and this is a needed area of research. It is apparent, however, that diseases do weaken and kill sweetgrass in cultivation, but sweetgrass in its natural habitat seems to be tolerant to diseases, and the hostility of the habitat may make it very inhospitable for these disease agents to express themselves.

Fertility Responses

On the ocean dunes, as the soil tests showed, there are a plethora of nutrients and many in excess. All plants require 19 essential nutrients including the macronutrients: nitrogen, phosphorus, potassium, calcium, magnesium, sulfur, carbon, hydrogen, and oxygen and the micronutrients including: iron, boron, chlorine, sodium, manganese, zinc, copper, molybdenum, silicon, cobalt, and nickel. The macronutrients are needed in major amounts and the micronutrients are needed in minor amounts. All elements are needed by the plant in specific amounts. All of these nutrients are considered essential because the plant cannot complete its full life cycle without each one. Sometimes when the supply of an element is excessive in the soil, it changes the pH of the soil and that in turn may cause an element to be more or less available. In some instances these changes can cause an element to become highly available and toxic. Nitrogen is the most important nutrient for plant growth. It is an essential component of all proteins. In its natural habitat, sweetgrass is never fertilized "by man" and depends on the environment to provide its essential needs. Somehow, sweetgrass flourishes in the wild without fertilization but in cultivation, sweetgrass usually may receive some fertilizer to "help" it grow. In some commercial areas such as parks and golf course, sweetgrass may be inundated with fertilizers and agrochemicals which will totally change its character temporarily. Sweetgrass in cultivation usually is not as desirable to basketmakers as the wild type. Overfertilization of sweetgrass does "change" the way the plant looks and feels and its value in basketry. Excessive fertilization with nitrogen can turn sweetgrass into what I call "Frankengrass" like the freak in Mary Shelley's "Frankenstein" novel. The spaghetti-tight leaves of sweetgrass will unfurl with high fertilizer, rending them more like a flat-bladed lawn grass. When harvested, this type of grass breaks apart in the hands of the puller, and it just is totally unusable in basketry that demands a tight coiled grass blade. Sweetgrass that is overfertilized is not permanently changed, and in time perhaps, in a few months, the leaves will again become tight and coiled once again after the fertilizer is leached from the soil. Fertilization is essential and important in the first year of establishment. Fertilization is essential during the greenhouse production of transplants and during their first year in the field (see discussion in Seedling Culture section). However, all fertilizer should be withheld permanently from sweetgrass destined for basketry beginning in the second year in the field and for all years to follow (unless some strange abnormality arises because some essential nutrient is present in excess or severely deficient) to cause the plant to become stiffer and coarser.

Chapter 15
Seedling Cultural Practices

Clump Separation to Produce Sweetgrass Plugs

In 1989, I started to grow sweetgrass at CREC and the only way I knew how to start new fields was to dig "mother plants" from the ocean front and separate the clumps into smaller separations or "mini clumps" and plant those (Fig. 15.1). At that time, we had no idea how to germinate sweetgrass and so we depended on digging sweetgrass mother plants from Kiawah and LSSI Islands for our inland cultivation fields. This method, as you can imagine, was an inferior approach since it depended the scarce supply of sweetgrass in its natural habitat. To produce sweetgrass, plantlets for acres of inland cultivation would require hundreds of mother plants. In the earlier days, we had a limited supply of sweetgrass mother plants so our first planting at CREC was made with separations that were about "dime"- and "nickel"-sized "mini clump" plantlets. From one large mother plant, separations can produce up to 65 mini clumps with about 4 leaves per mini clump. If the roots of the separations were excessively long, I would root prune so they fit into their containers better. Of the 50 "mini clump" plantlets planted, only about 35% survived the first summer. Although I have tried planting "mini clump" separations in spring, summer, and fall, I'd prefer to dig my mother clumps in late October, make the separations, and grow the plantlets through the winter in a greenhouse until mid-February for an early March field planting.

The use of clump separations only has utility in cultivated fields though, if it is necessary to move plants and/or to make new fields by using older fields. In some fields that I wished to transplant or move to new locations, I realized that the size of the plantlets that were bare root transplanted from the old location to new location must be much larger than the "dime and nickel" size used. Bare root transplanting is simply splitting the original mother plant and replanting soon afterward in the field. This process avoids the growing of mini clumps in the greenhouse and is typical of how gardeners separate perennial flower species in the home garden. I have found that separating mother plants into "fist"-size separations increased survival to almost 100% (Fig. 15.2 a, b). It was obvious that bare root plantlets must be large or else plant death will be very high. This system works well for established inland cultivated

Robert J. Dufault, *Stalking the Wild Sweetgrass*, SpringerBriefs in Plant Science, DOI 10.1007/978-1-4614-5903-3_15, © Robert J. Dufault 2013

Fig. 15.1 A large clump of sweetgrass dug in the natural habitat on Kiawah Island, SC

Fig. 15.2 A clump being separated into "fist"-size plantlets (**a** *left*) and final plantlet clump size compared to a ball point pen (**b** *right*) and ready for bare root field planting

sweetgrass but to dig plants in their natural habitat for inland planting makes the supply situation even direr since the supply of mother plants is very limited.

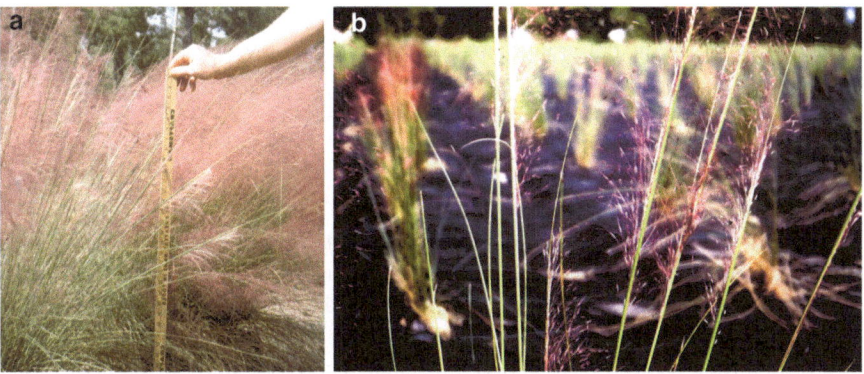

Fig. 15.3 Sweetgrass in full bloom with immature seed in early October (**a** *left*); close-up of individual seed heads within the flowering plume (**b** *right*)

Seed Germination and Seedling Culture

The greatest bottleneck to acquire planting material for new inland cultivation sites was the lack of mother plants that could be sacrificed for vegetative plantlets. In the early 1990s, I gathered seed from sweetgrass plants and tried to germinate them. To my dismay, my germination rate was less than 1%. Sweetgrass in the Charleston area sends out its beautiful mauve seed head "blooms" in mid- to late September. The plants will stay extremely showy for about 1 month and then start to turn mauve brown and still remain attractive yet not as colorful until December or later.

In my first attempt to germinate sweetgrass, I gathered seed in late October but had poor germination results (Fig. 15.3 a, b). It always had been an enigma how sweetgrass propagates itself on the dunes. Sweetgrass does not send out runners like many grasses and it is strongly felt that new sweetgrass plants in its natural habitat are "started" by volunteer seeds. After some thought over time, I realized that if that was true, at some point in the seed development, they must fall off the plant and get imbedded into the sand soils and germinate when the conditions are "right." The secret to seed harvest was simple. I watched plants through the late fall and early winter and I noticed when the seed started to naturally fall off of the plants. The seed of some grass species is immature or underdeveloped after flowering and needs some time to ripen and mature on the plant. This is the case with sweetgrass.

For germination success to be ideal, sweetgrass seed needs to stay attached to the plant to develop and mature. I determined that the ideal time to collect sweetgrass seed at proper level of maturity is between Thanksgiving and Christmas holidays. This is the normal time when the seed naturally "shatters" from the plant and disperses into nature in its attempt to ultimately germinate. I have, over the years and on numerous occasions, harvested sweetgrass seed at this time and the germination rate is always better than 90%. So this mystery was solved and without the bottleneck of digging mothers from the natural habitat, new inland plantings could be started from seed using cultural practices similar to growing common vegetable plants. So what are the cultural practices needed to grow sweetgrass?

When to Seed

Sweetgrass is a warm season perennial grass and it will grow best in warm conditions in the field. It loves the heat but does grow throughout the winter in cold conditions. Planting in spring should occur about late March to early April in Charleston so the transplants can immediately begin very active growth. It takes about 12 weeks to grow a thrifty vigorous sweetgrass transplant. Therefore, mid-December is good time to start spring sweetgrass. The seed should be harvested as mentioned above. If planting will not occur until a later time, the seed can be refrigerated. Sweetgrass seed will remain viable for years if it is stored soon after harvesting in the freezer in a ziplock storage bag.

Containers

What type of container or cell pack should you use to grow sweetgrass (Figs. 15.4 and 15.5)? The answer is really any container can be used as long as it has good drainage and suits the grower. However, if the sweetgrass is destined to be planted into an area that will not receive any additional water and will depend on rainfall solely (ocean dune plantings and commercial sites without irrigation resources), I would suggest using a "nursery tray" with dimensions about 5 in. deep by 2.5 in. top diameter tapering to about 2 in. diameter at the bottom (Fig. 15.4).

Sweetgrass cannot be directly seeded into the field soil. Sweetgrass is a very delicate grass that does not compete well at all with weeds and can easily be lost. I have noticed over the years though that sweetgrass will volunteer seed in cultivated locations very rarely, but it is not a "weedy" plant that "volunteer seeds" everywhere. Even though a single clump of sweetgrass can produce thousands of seed, the plant does not reseed itself easily.

Seeding the Media

Starting sweetgrass seedlings is quite easy and uses a similar approach as starting vegetable crop seedlings. Once you decide on the container you will use, fill the containers with a very light seedling mix like a vermiculite/peat/perlite mixture. Garden soil or potting mix should not be used since you need a media that is very light with good drainage. Seed in the refrigerator/cold storage needs to be cleaned of stems and other seed head trash. An easy way to this is to use a kitchen colander. Brush the "clumps" of seeds on the colander bottom, almost like softly grating

Fig. 15.4 Deep nursery tubes are better for planting in very dry locations such as the beach because roots go much deeper into the sands. These cells were used for the Grand Strand planting

cheese, and the individual seeds should come through the holes and forms a small pile. If the colander has big holes, the seed should fall through the holes. You may want to shake the seed/straw mass to free seed as you go along. This is why it is important to collect a large amount of seed since some seed refuses to come off the straw. If this procedure is not successful, you can still plant straw attached to the seed, but it isn't as easy to cover the straw. However, the colander trick should work well and the seed should be singulated now and easier to plant. When seed is cleaned, you will be spreading the seed over the entire top of the container in a fashion like thinly spreading confectionary sugar on cookie. Then, brush off any on the container edges back into the planting media. You are planting a "community" of plants since in nature, a sweetgrass plant is a congregation of many plants in one clump similar to Lirope (monkey grass) or pampas grass (Fig. 15.6a, b). If you dissect a clump of sweetgrass, you can easily separate small plantlets with each plantlet having 5–10 leaves each.

When seeding sweetgrass, planting a single seed in each cell is not correct since you will be producing a very slender, scrawny plant. In each square inch of planted surface, you should have about 20 seeds to successfully produce a vigorous mini sweetgrass clump (Fig. 15.6). After seeding, sprinkle a thin cover (1/8″) of planting

Fig. 15.5 Four in. by 4 in. cell packs can be used to grow very large-size plants. These cells were used at Folly Beach in our first beach planting, but longer nursery tubes are preferred

Fig. 15.6 A clump of sweetgrass is really a community of many plants growing side by side. This clump is composed of about 50 plants (**a** *left*); a "mini clump" of seeded sweetgrass ready for planting in the field (**b** *right*)

media over the tops of the seed containers, water very gently with a fine spray, cover with a sheet of saran wrap to retain moisture, and put in a greenhouse or in a sunny window and wait. Within a few days, the sweetgrass should emerge and the saran

Fig. 15.7 Newly germinated sweetgrass seedlings after planting in a long seedling flat. The flat now looks like a "mini sweetgrass lawn"

wrap taken off. If you have been successful, you should see "mini sweetgrass lawns" as the result of broadcasting the seed in each container (Fig. 15.7).

Seedling Fertilization

Once the seedlings have emerged and are about one inch tall, I would use some basic soluble fertilizer like 20–20–20 sold in store sometimes as "tomato fertilizer" (20–20–20 is 20% nitrogen–20% P_2O_5–20 K_2O)(Fig. 15.8). Mix about one tablespoon to a gallon of water and fertilize every 2 weeks with this solution. Whenever the seedlings get too tall and leggy (stretched), they can be given a "haircut" by cutting off a few inches to maintain the plant height to about 6 in. The problem with longer leaves is that they may "fall over" their neighbors and create more competition. If light levels are low, you will have trouble keeping the plants from stretching so put them in the strongest light available. If the seedlings continue to grow very tall and lanky, stop fertilizing at that rate and only apply fertilizer when the leaves turn "pale green."

Fig. 15.8 During seedling
growth and the first year in
field, fertilizer is very
important to encourage
vigorous plant growth. In
2nd year, however, no
more fertilizer is used for
sweetgrass destined for
basket use

Seedling Watering

Mature sweetgrass is a drought-tolerant plant and can withstand moderate water stress. In the seedling stage, however, sweetgrass requires regular watering. I would tend to water LESS than MORE in the seedling stage before planting outdoors. It is difficult to predict when to water since each environment may be drier or wetter. The best way to tell the water status is stick your finger in the media and then pull it out. If it feels cool and somewhat damp, the media has enough moisture and do not water until it feels "drier." The media used to plant sweetgrass should be very light if you followed the instructions above. If potting or garden soil have been used to grow seedling sweetgrass, these media hold too much water and the roots of sweetgrass may rot if standing in water for a long period of time. Using the finger method above though should be a fail-safe method to prevent overwatering.

Seedling Hardening

Before sweetgrass plantings are planted in the field, the plants should be moved outdoors for at least a week before planting. Place the plants in a sunny location when hard freezes have passed (mid-March in Charleston). "Hardening" exposes

Fig. 15.9 A quality sweetgrass plant is a stocky, vigorous plant with many "threads" (leaf blades). The leaves should have a wire-like texture and be about 6–8 in. long with a finely knitted root-ball about 5–6 in. long. The color of the leaves should be emerald green and not lime or pale green. The roots should be white or light tan but not brown. In many commercial plantings, the leaves act as "handles" to pick up plants and carry them to their planting holes. New leaves will rapidly replace or add to these seedling leaves. The greater the leaf strength insures that survival rate will be better as well

the plant to harsher weather and helps them toughen up before they are moved to their permanent position. During this time, watering should be reduced and fertilization should be withheld until the day BEFORE transplanting. Fertilizing the day before and the day of planting will give the plant a boost to help in the recovery after transplanting.

What Is a "Quality" Sweetgrass Transplant?

How do you know if you have produced a quality sweetgrass transplant (Fig. 15.9)? What are the aspects of the plants that are good indicators that you have been a successful grower? The following outlines my criteria for a quality transplant.

What Are the Problems in Growing Sweetgrass Transplants?

Fertility

From experience, one aspect of growing seedlings that was very problematic was excess fertilization. I would avoid using a timed-release fertilizer simply because they usually are used in excess and release too much fertilizer when you do not need it. Timed-release fertilizer releases fertilizer in response to temperature. As temperature increases, more fertilizer is released. If used very sparingly, timed-release fertilizer is acceptable yet most people think that "if a little is good, more is better." Sweetgrass is not a gross feeder of fertilizer and only mild applications are necessary to sustain adequate growth. Also, if plants are to be planted in their natural habitat as part of some restoration project, the seedlings need to be hardened by the time they get to the field and it is imperative that the hardening process is used as discussed above.

Low Light

Growing sweetgrass in low light, shady locations induces spindly growth and weak leaves. Sweetgrass should receive as much light as possible and I would suggest at least 8 h of strong light each day. Low light will cause the plants to stretch, then fall over and essentially become a "hanging plant" and potentially "fall" on their neighbors and shade them. If they do become spindly, cut them back to about 6 in. to help them "stand up."

Seeding Only a Few Seeds per Container

Seeding the containers with too few seeds per container will produce a stringy, weak seedling. These spindly seedlings are very difficult to pull from their containers at field planting and may be seriously damaged and set back in development. Weak seedlings will grow much slower, require more time to eventually form large clumps, and may require more years to reach harvestable size.

Overseeding

Seeding the containers very heavily can cause too much competition in small containers and eventually the seedlings will decline and many will die. If this happens, thin out the excess seedlings to reduce competition.

Greenhouse Growth Period Was Too Short

If not enough growth time was given from seeding to planting, the resulting seedling may be very weak and spindly. Reducing the growth period of seedlings and planting weak transplants will just require more time for the plants to reach harvestable size in the field. On the other hand, old sweetgrass seedlings that cannot be planted in the field for some reason can be kept for a year in larger containers if necessary (not advised though). I have kept seedlings in their containers outside throughout the winter until the spring planting without loss. However, these seedling root systems became very root bound without adequate fertilization. Watering must be religiously monitored if seedlings are carried over through the winter.

Excessively Long Leaf Production

In low light conditions and/or with excessive fertilization, the seedlings may produce leaves over 12 in. in length. These leaves will eventually shade other plants out and deflect watering that needs to soak the root-balls. In that case, the excessive growth can be trimmed periodically to about 6 in. to allow light and fresh air to penetrate the base of the plants, and this helps keep the plant healthy. When seedlings get large but cannot be field planted, thorough watering is necessary and essential to reach and saturate the root-ball.

Chapter 16
Field Production Practices

There are many aspects of growing sweetgrass successfully that need to be discussed as follows:

Timing of Spring Planting in the Field

Sweetgrass loves the heat and grows rapidly into the summer months as most grass species do. Even though sweetgrass is a heat lover, spring planting needs to be done early to avoid the really hot weather that arrives in late May and thereafter. The major reason is most inland fields planted to sweetgrass depend on natural rainfall for irrigation purposes, and heat stress after transplanting can surely kill the struggling seedlings. The best time to get sweetgrass in the field is after the last spring frost which occurs approximately March 21st in the Charleston area. Early planting also avoids those times of massive weed invasions usually occurring with the heat in late May and throughout the summer.

Soil Conditions

The soil conditions common in the sweetgrass natural habitat were explained before. Ideally, a sandy soil would be ideal that is low in organic matter. In the production of sweetgrass, I would not recommend using any organic matter additions like compost and/or animal manure. I would not enrich the soil with any slow release amendments which would cause the sweetgrass to grow luxuriously and very succulent. If sweetgrass is to be used strictly for ornamental value, however, I would use organic matter amendments, but in those cases, the leaves will not be harvested and all the grower cares about is "pretty" and not functional basket construction material.

Robert J. Dufault, *Stalking the Wild Sweetgrass*, SpringerBriefs in Plant Science, DOI 10.1007/978-1-4614-5903-3_16, © Robert J. Dufault 2013

Fertilizer Additions in the Field

In the first year only, the newly transplanted sweetgrass should be given a "shot" of some soluble, readily available fertilizer that will last the plant the first growing season. I would suggest using a tablespoon of a timed-release fertilizer in this situation that is placed in the bottom of the planting hole at planting time. I would use a complete fertilizer such as 20–20–20 or another fertilizer with equal numbers to supply nitrogen, phosphorus, and potassium, respectively. During the growing season, if the sweetgrass seems not to be growing vigorously, I would prepare a foliar drench of 20–20–20 again, probably one tablespoon per gallon, and pour the solution on top of the plant. If the sweetgrass turns a very pale green, I would add Epsom salts from the drug store at about 2 tablespoons per gallon as a foliar drench. Epsom salts is essentially most magnesium which is a major element and component of the chlorophyll molecule that is involved in photosynthesis (makes sugars for the plant metabolism) occurring in the plant.

Watering

One of the greatest killers of newly planted sweetgrass is lack of water after planting. Sweetgrass, like so many other transplanted crops, is very vulnerable to water stress until the transplant roots finally "knit" well into the soil and grow deeply to access soil water. If irrigation is available, an inch of water weekly should be applied to the newly planted seedlings if natural rainfall does not occur. Plants that will not be irrigated and will depend on natural rainfall should be planted into a slurry of Terra-Sorb as described above in the Folly Beach planting operation. Use of Terra-Sorb is mandatory on beachfront planting sites and really for all nonirrigated situations. This hydroscopic gel mass will retain water near the roots of the plant for months if not years and help it survive the critical time period after transplanting. By midsummer after the plants have finally established well and are actively growing, watering on an irrigation schedule can be cut back since it would be desirable for the plants to send deep roots into the soil to search out water. Excessive watering causes roots to stay superficial and makes the plants "lazy." In the 2nd year and thereafter, sweetgrass used for basketry needs to be exposed to environmental stress to make the leaves very stiff and tough. Excessive watering will make the plants very succulent and shatter if pulled for basket needs.

"Benign Neglect"

The practice of withdrawing nutrients and water and exposing the sweetgrass to environmental stresses is what I call "benign neglect." This is the process of toughening up the plant to grow strong and resilient to stresses which cause the leaves to become very coarse but ideal for basket uses. In a way, benign neglect is very much like hardening, but benign neglect is used on mature plants growing in the field and not seedlings. Sweetgrass used for strictly ornamental use should be fertilized and watered but not for plants grown for basketmaking. This is one of the reasons why cultivated sweetgrass is not liked by basketmakers. It usually has been fertilized, is succulent, and breaks easily. Plantings installed at parks, golf courses, residential neighborhoods, shopping centers, etc. are usually mixed plantings with other plants and may receive nutrients and frequent, sometimes daily, irrigation. The sweetgrass in many cases turns into "Frankengrass" which is low quality for basketry.

Weed Control

Sweetgrass planted in small gardens will usually be hand weeded. Sweetgrass planted on the ocean front will also most likely "be on their own" after planting without any human intervention. Sweetgrass planted in 2005 at Folly Beach has never received any attention from humans since the day they were planted, and they have established very well and competed with the few weeds on the dunes also very well. Sweetgrass planted in large inland fields, however, requires farming technology to keep weed infestations manageable. In those cases, agrochemicals or herbicides would be strongly advised for use. Over the years, I have experimented with common grass/corn herbicides that control mostly broadleaf weeds with varying success. The following herbicides have been used successfully:

Basagran postemergence spray applied over the top of sweetgrass plants to kill broadleaf weeds.

Surflan preemergence spray applied in spring to bare ground between the rows to prevent weed seed germination.

Atrazine preemergence spray applied in spring to bare ground between rows to prevent weed seed germination.

Accent applied as a directed postemergence spray to the weeds but avoid spraying the sweetgrass.

2, 4-D amine (*Weedone*) applied as a directed postemergence spray to the weeds but avoid spraying the sweetgrass.

Paraquat (*Gramozone*) applied as a directed postemergence spray to the weeds but avoid spraying the sweetgrass; never apply this herbicide on the sweetgrass or it will kill!

Herbicides that have been used which seriously damaged sweetgrass should be avoided and include the following: Lorox, Sencor or Lexone, and Poast.

If large acreage is planted, it is highly recommended that the rows be spaced apart very wide (6 feet minimum). The wide spacing will allow a tractor to cultivate in the middle of the rows. At some point, however, the sweetgrass will become very large and "fountain over" the ground, making close cultivation impossible and destructive to the leaves. In those cases, hand pulling may be the only prerogative. If weed control "gets out of hand" though and the weeds overtake a planting, I would suggest using winter renovation. In January, the sweetgrass is mowed down to about 10 in. in height and the dead leaves removed from the fields; then a preemergence herbicide as recommended above can be applied to the soil to prevent weed seed germination in early spring. During winter renovation, sweetgrass should never be burned because the growing points may be destroyed and the plant seriously hurt.

Light, Shade, and Competition

Sweetgrass does not tolerate moderate to heavy shade. If sweetgrass is to be planted for basketry, the field must be exposed to full sun for the majority of daylight hours. If sweetgrass is planted near other ornamentals, it will not do well and may remain very small and spindly, and if shade is too intense, it will die. If planted back to a natural setting on the ocean dunes, the situation is different. In its natural habitat, sweetgrass thrives if growing near wax myrtle plants. The shade provided by the bushes tends to reduce the heat load and make the sweetgrass grow more luxuriously. It could be that wax myrtle naturally occupies the lowest lying areas of the dunes and that water collects there as well. Sweetgrass, in turn, may have a greater supply of water in those same areas, hence the longer leaf growth. Plantings made at Folly Beach were placed under walkover bridges that carry people from the parking lots over the dunes to the beach. The light shade in those areas which are usually mini ravines was excellent place for the best-quality sweetgrass that was planted at the beach. It must be remembered though that the environment at the beach is so very harsh, hot, and dry and with salt breezes; the refuge provided by the walkovers very much enhanced the growth of sweetgrass.

Plant Spacing in Fields Planted for Basketmakers' Use

My recommendation for planting sweetgrass production fields is to plant them 6 feet apart in rows, with rows spaced at least 6 feet apart. This recommendation comes from viewing the huge sweetgrass fields at LSSI. It was very obvious that sweetgrass naturally determines its own best spacing themselves. It was remarkable how the plants seemed to be evenly spaced almost like someone planted them at LSSI. The plants all appeared to be about 5 feet apart in all directions and each plant tended to fountain over the ground. These fountains, however, made wonderful cool moist habitats for rattlesnakes which were plentiful on LSSI and Kiawah. The wide spacing I recommended allows enough room for a basketmaker (1) to walk around the plant to judge any dangers and (2) to allow the plants to grow long luxurious leaves without excessive competition. I have planted sweetgrass one foot apart in rows 6 feet apart, but for the sake of varmint control, it is best to give the plant a wide berth so the base of the plants is visible. The recommended plant spacing above requires about 1,450 plants per acre (an acre is about 200 feet long by 200 feet wide).

Timing of Fall Planting in the Field

I have tried planting sweetgrass late in the fall to see if this new approach would work, and surprisingly, it worked beautifully. I always have been hesitant to try this technique because sweetgrass loves heat and grows best in the warm summer months, but sweetgrass does continue to grow throughout the winter. It has been my experience that late-fall-renovated plants will continue to initiate new leaves throughout the winter months, producing their new green leaves sprouting from the centers of the clumps. One of the advantages of fall planting is that it is cooler and easier on the people planting and the weed problems are very easy to manage since only weak winter weeds are common and not the vigorous sometimes rampant summer weeds. Also, fall-planted sweetgrass does manage to grow significantly during the winter months and a quarter-sized diameter seedling may grow to about 3 in. in diameter by the following spring if planted in an ideal location. If fall planting is desired, I would suggest seeding in July for field planting in early November using all the techniques discussed above. Sweetgrass is quite tolerant of frosts and only the most severe frosts will burn sweetgrass. It also should be remembered that the growing points of seedling sweetgrass are very close to the ground which is warm and the soil actually gives off some heat even on very cold nights. Other plants that are tall and totally exposed to the freezes will suffer greatly compared to the very low growing sweetgrass seedlings. Another facet of this planting method is if field planting, for whatever reason, cannot be made in the fall, the sweetgrass seedlings can be left outdoors and they will survive the winter until planting the following spring. Of course, the seedlings should be watered and watched to assure they do not dry out.

Growth Practices in the First Year After Planting

Many of the aspects of production during the first year have been expanded on above. It must be remembered in the first year that extra care must be taken with sweetgrass to get it established. Sweetgrass planted in inland locations needs special care. Sweetgrass must be watched carefully for (1) invasion of weeds, (2) drying out or drowning, (3) and damage by insects, (4) making sure the plants are adequately fertilized to encourage rapid, vigorous growth. A light side-dressing of any complete fertilizer (10–10–10) around the base of the plants every 6 weeks during the first summer will help increase plant girth and clump diameter. The leaves will lose *temporarily* the typical sweetgrass tight wire-like appearance and the leaves may even open up similar to blades of grass. However, the first year is the only year that sweetgrass is fertilized for use in baskets. Fertilizer will "run out" in a few months and the leaves will "revert" back to their natural condition as tight rounded leaves. However, sweetgrass planted back into its natural habitat on the ocean dunes will be on its own without any care after the initial planting. It would be suggested that basketmakers who ultimately harvest plants planted on the ocean dunes try to clean up the plants from any weeds that compete with the plants after they harvest the plants.

Growth Practices in the Second Year After Planting and Beyond

Continue to use *benign neglect*. Beginning in the second growing season, the clumps should be 6–8 in. in diameter and this is the time when they need to be grown to produce quality sweetgrass in the third year. Sweetgrass growing along the dunes is subjected to extremes of heat/water/nutrient/salt stress which produces very tough leaves. Therefore, in cultivation, water and nutrients should be avoided beginning in the second year after planting. Fertilization and copious watering will cause the leaves to become very succulent and potential "open up," and this makes pulling at harvesting impossible and the leaves may break. The grower needs to watch over the plants and make sure they are not shaded or infested with fire ants or the weeds totally overwhelm the planting. Weeds, shade, and fire ants are the biggest deterrents to quality sweetgrass production. Remember to think about sweetgrass's natural habitat. On the dunes in a desert-like environment where they have evolved to tolerate extreme stresses including salt and sand blasting from the wind, drought stress, and lack of fertilizer and any human help, they do fine. Attempting to provide sweetgrass with overattentiveness really is a major problem with home horticulturists and farmers who are used to growing food crops including too much fertilizer and water.

Changing Sweetgrass into "Frankengrass"

Sweetgrass growing in its natural habitat on the dunes along the ocean is exposed to extremes of weather conditions, and it is a very tough plant to thrive in such hostile conditions. These plants get no attention whatsoever from "caretakers" and are incredible survivors. The beach is a desert condition. One of the beauties of their natural environment is the lack of natural pests such as insects and disease agents. However, in inland cultivation, plantings are exposed to different agents that affect growth, such as insects, disease agents, richer soils with higher water-holding capacities, weed species, and agrochemicals. All of these agents can "change" how the plant grows and, in some cases, can render the sweetgrass useless for basketry and change it into "Frankengrass." For example, many of the plantings made on golf course receive nutrients, water, and agrochemicals which may increase the inferiority of the sweetgrass leaves such as increasing their brittleness. If sweetgrass does revert to Frankengrass, it can be changed back to quality sweetgrass in about 6 months if benign neglect is used.

Harvesting

Sweetgrass planted on ocean dunes will require three full growing seasons for plants to grow large enough to be harvested. The stress on the ocean front slows the growth and development down significantly compared to sweetgrass planted in inland fields. In the latter's case, it will take about two full growing seasons to get the sweetgrass clumps large enough for harvest to begin. Harvest time in Coastal SC begins in the spring, but the leaves grow longer and longer as summer progressed. Sweetgrass flowers from mid-September in this region, but seed stalks begin to form in the bases of the plants in late August through mid-September, and their presence makes the sweetgrass low quality and unusable.

Yearly Renovation

Over the years, I have debated the advantages, need, and practices of the removal of dead leaves in winter. Usually the leaves do turn brown and die, and unless removed, they eventually rot. Some people in sweetgrass community feel it harms the plant to cut them back, and they resist this practice. Some feel that combing the plants, as you would brush your hair, may help in removal of dead leaves which indeed will help the situation versus doing nothing. On the large scale though, this would be too laborious. I encourage clump renovation each year, but it must be handled in a very specific way. As stated before, a clump of sweetgrass is a community of many individual plants growing jammed together in a clump. Over time, individual plants within the clump die and decay. When sweetgrass is harvested, much time is spent removing and throwing away these dead leaves from the handfuls pulled. My recommended method of renovating sweetgrass is to trim back the plants in the dead of winter, preferably in January, to about 8 in. tall. Never mow sweetgrass to the ground and absolutely NEVER burn sweetgrass!! The growing points of the plant are very close to the surface of the ground and burning will melt and destroy the tender growing points. Excessively short mowing can damage growing points as well. Even though sweetgrass is a warm-season grass, it continues to slowly grow in winter but very slowly. It is not unusual to notice soon after blunt cutting the clumps the emergence of new leaves even in January.

The advantages to trimming the sweetgrass are as follows:

Removes decaying plant material and reduces sorting in future harvests. If dead leaf material remains, it harbors insects and the leaves will decay and invite potential disease organisms. Unkempt sweetgrass clumps can also become habitats for critters in inland production fields.

Allows sunlight to get into the heart of the plants. This helps dry up the hearts and reduces holding moisture in these dead parts that may lessen disease.

Allows new plantlets to thrive without the competition of those that died.

Longevity and Decline of Sweetgrass Plants in Cultivation

Interestingly, sweetgrass in its natural habitat on the beach dunes appears to live for decades, yet in cultivation, they are considered short-lived perennials. In observations of sweetgrass plantings in the municipal setting, plantings live about 5–8 years and decline. In some of these situations, I have observed no yearly renovation and the dead leaves are the major visual feature. In the cultivated setting, there are greater dangers to disease and insects that can cause serious decline and death in a few years. In my own plantings, the sweetgrass grew beautifully for about 5 years then declined and the clumps became very spindly. It is not considered a failure that a "perennial" died within 5 years since this is customary with some perennials. All plants are terminal although some perennial species have a much shorter lifespan. The natural habitat for sweetgrass is so incredibly harsh yet that environment seems to sustain a long lifespan for wild sweetgrass. I do not know anyone who has followed the lifespan of sweetgrass on the ocean dunes, but with plantings made at Folly Beach and the Grand Strand, we will gain this knowledge as the years pass.

Chapter 17
Concluding Thoughts

What Mysteries Are Left to Unravel?

There are many unknowns still about what makes sweetgrass "tick" but we know a lot more now since 1988!

We truly do not know the exact longevity of sweetgrass growing in natural habitat and we do not understand why sweetgrass declines so rapidly in cultivation in comparison to their natural habitats. There is a "disconnect" between the harshness of the natural environment which apparently increases its lifespan in contrast to the toxicity of the inland locations that apparently terminates sweetgrass otherwise long life prematurely. Research is needed to help understand this enigma.

In the inland production fields, what are the plant diseases that destroy sweetgrass? Can they be controlled with agrochemicals or other cultural practices? Research is needed here also.

Why are fire ants drawn to these plants? There is a strong attraction of fire ants to sweetgrass that needs research. It is common in a production field to find, after a few years, a significant percentage of plants are habitats for fire ants. These insect pests can be controlled with insecticides, yet they are a tenacious insect pest, many people are highly allergic to a single fire ant sting, and residues of potent insecticides needed to kill fire ants are a safety hazard for people harvesting the sweetgrass. Production fields of sweetgrass need to be clear of these pests.

What is the true genetic diversity of the sweetgrass germplasm that grows in existing natural habitats? When we talk about sweetgrass, we are assuming that only one genetic type exists and it is a monoculture. We do not know the genetic variation that exists in the wild plant population. Most of the cultivated plants are "descendants" from those removed originally from the dunes and then propagated either vegetatively or by seed. In either case, the parentage is very narrow compared to the biodiversity that still exists in the wild populations. I assume that there is a wide variety of biotypes in the wild with different traits of hardiness, fiber strength, length, and other phenotypic characteristics carried in the plants' genetic makeup. In the future, a DNA profile of sweetgrass in diverse locations throughout the

Robert J. Dufault, *Stalking the Wild Sweetgrass*, SpringerBriefs in Plant Science, 115
DOI 10.1007/978-1-4614-5903-3_17, © Robert J. Dufault 2013

Southeast would provide proof that within *Muhlenbergia sericea,* there is a diversity of genetic material, and potentially, ecotypes could be selected and breed for greater length, fiber strength, fiber girth, disease and insect resistance, and longevity. Dr. Gustafson at the College of Charleston, Dept. of Biology, reported that he has conducted genetic research on *M. sericea* from Texas and South Carolina (Gustafson et al. 2008). He found that the plants were genetically different and he identified ecotype variation between the two populations. So apparently, the wild sweetgrass population is a heterogeneous population with many different genetic resources still left intact. It is possible in the future that cultivated varieties could be bred to possess outstanding basket qualities.

What are the best inland cultivation fertility practices to increase fiber strength and thread length? Should we try to replicate the soil conditions in inland fields, based on natural habitat soil tests to increase the quality of cultivated sweetgrass? Do these soil tests provide clues of the soil environment that may best enhance sweetgrass cultivation in an inland setting? Soil test results from the ocean dune habitat are very different from a good garden soil used inland for crop production, but is it a requirement to change the soil fertility to mimic the natural habitat, or is it a fluke and can be ignored? Does the extreme fertility situation outlined in soil test results from natural habitats actually increase sweetgrass longevity? These are all highly researchable topics.

Chapter 18
Afterthoughts

Many years have elapsed since that day long ago when I attended that meeting of many individuals from various state institutions and colleges trying to take the task up solving the supply dilemma. Over the course of more than two decades, all kinds of attempts have been made to STALK THE WILD SWEETGRASS and TAME IT, and many have failed. Yet in those failures, we learned a lot about this still elusive plant. We started from taking this plant from its ancestral home and trying to "make it work" in environment not really suited to its welfare. After two decades, our efforts have taken it back to the beach as part of the habitat restoration of the Army Corps of Engineers which I feel is the greatest accomplishment of the two decades of work. Although we can grow it "OK" inland for a while, the sweetgrass' return to the coastal habitats, will, in years to come, spread again on state land that is protected from developers forever (it is hoped). On May 15, 2010, I invited a small group of basketmakers to evaluate the quality of the sweetgrass and harvest as well, before the sweetgrass fields would be open to all basketmakers for harvest (Fig. 18.1). They commented that the plants produced good-quality sweetgrass but the plants could use some "combing" to rid the plants of dead leaves. They stated, as I too believed, that removal the dead leaves is very important to the health of the plants. Furthermore, not removing the dead material causes the plants to decline at a faster rate and "combing" the plants tends to invigorate the plants and allows the new leaves to emerge faster. As these basketmakers harvested, they also removed much of the dead leaves. It was apparent that the Folly sweetgrass story is now just beginning and that its survival in years to come will be at the benevolence of the harvesters and the manner in which they help maintain the health of these plantings.

In this respect now, I feel I can say that this journey of mine with sweetgrass has a "happy ending" or a happy beginning by returning sweetgrass to the beach through the magnificent help of the Army Corps of Engineers. This is the best result I could have ever dreamed of.

It is important to note that the Sweetgrass Cultural Arts Festival Association (SCAFA) is the premier local organization with a mission to preserve the heritage, legacy, and traditions of the Gullah Geechee culture and their sweetgrass

Robert J. Dufault, *Stalking the Wild Sweetgrass*, SpringerBriefs in Plant Science, DOI 10.1007/978-1-4614-5903-3_18, © Robert J. Dufault 2013

Fig. 18.1 On May 15, 2010, a pilot group of basketmakers evaluated and harvested the Folly sweetgrass for the first time

basketmaking art form. They also promote the protection of sweetgrass natural habitats and its environment from destruction by utilizing SCAFA as a venue to bring focus and attention to the declining natural habitat locations where basketmakers harvest the sweetgrass needed to produce their art form. SCAFA also establishes collaborative partnerships with state, local, and national organizations/agencies to support our efforts to protect and preserve the Gullah Geechee culture and the sweetgrass basket art form. And lastly, SCAFA is totally inclusive of all sweetgrass basketmakers in their efforts pursuit to protect and preserve the culture and art form. Over the years, SCAFA has been involved and informed of the natural habitat restoration activities at Folly Beach. They are working with the local government on Folly Beach to develop ways of harvesting sweetgrass plantings in an organized and cooperative manner. SCAFA has also identified at least 4 other areas in the Charleston area that sweetgrass has been planted and can and are being harvested through their organization.

References

Charleston Gazette and Advertiser (1791) South Caroliniana Library, Columbia, SC

Gustafson D, Peterson P (2007) Re-examination of *Muhlenbergia capillaris*, *M. expansa*, and *M. sericea* (Poaceae: Muhlenbergiinae). J Bot Res Inst Texas 1:85–89

Gustafson DJ, Halfacre AC, Anderson RC (2008) Practical seed source selection for restoration projects in an urban setting: tallgrass prairie, serpentine barrens, and coastal habitat examples. Urban Habitats 5(1):7–18. Available at http://www.urbanhabitats.org

Joyner C (1984) Down by the Riverside: A South Carolina Slave Community. Univ. Illinois Press, Urbana, IL

Kurz H, Wagner K (1957) Tidal marshes of the Gulf and Atlantic Coasts of Northern Florida and Charleston, South Carolina. Florida State Univ. Studies, No. 24, The Florida State Univ., Tallahassee. p 168

Littlefield DC (1981) Rice and slaves: ethnicity and the slave trade in colonial South Carolina. Univ. Illinois Press, Urbana, IL

McKissick Museum (1988) Proceedings of the Sweetgrass Basket Conference, McKissick Museum, The University of South Carolina, Columbia, pp 79–84

Meyers R, Ewell J (1990) Ecosystems of Florida. Univ. Presses Florida, Gainesville

Middleton M (1990) From Conference to Birth of an Association. In: The Sweetgrass Heritage, Newsletter of the Mt. Pleasant Sweetgrass Basketmakers' Association, vol 1(1). p 3

Ohlandt K (1992) Where the sweetgrass grows: the restoration of maritime wet grassland, incorporating harvesting *Muhlenbergia filipes*. Thesis of The Univ. Georgia, Master of Landscape Architecture, Athens, Georgia. p 118

Peterson P (2003) Muhlenbergia Schreb: In: Barkworth ME, Capels KM, Long S, Peip MB (eds) Magnoliophyta: Commedlinidae (in part): Poaceae, Part 2. Flora of North America North Mexico, vol 25. Oxford University Press, New York, NY. pp 145–200

Pinson JN Jr, Batson WT (1971) The Status of *Muhlenbergia filipes* Curtis (Poaceae). J Elisha Mitchell Sci Soc 87:188–191

Rosengarten, D (1987) Row Upon Row: Sea Grass Baskets of the South Carolina Lowcountry. Mckissick Museum, Univ. South Carolina, Columbia

Vlach JM (1978) The Afro-American tradition in decorative arts. Cleveland Museum of Art, Cleveland, OH

Wexler M (1993) Sweet tradition. Nat Wildlife Magazine 31(3):38–41

Robert J. Dufault, *Stalking the Wild Sweetgrass*, SpringerBriefs in Plant Science, 119
DOI 10.1007/978-1-4614-5903-3, © Robert J. Dufault 2013